M. C. Cooke

Toilers in the sea

M. C. Cooke

Toilers in the sea

ISBN/EAN: 9783743343887

Manufactured in Europe, USA, Canada, Australia, Japa

Cover: Foto ©berggeist007 / pixelio.de

Manufactured and distributed by brebook publishing software (www.brebook.com)

M. C. Cooke

Toilers in the sea

TOILERS IN THE SEA.

BY

M. C. COOKE, M.A., LL.D.,
AUTHOR OF "FREAKS AND MARVELS OF PLANT LIFE," ETC.

PUBLISHED UNDER THE DIRECTION OF
THE COMMITTEE OF GENERAL LITERATURE AND EDUCATION
APPOINTED BY THE SOCIETY FOR PROMOTING
CHRISTIAN KNOWLEDGE.

LONDON:
SOCIETY FOR PROMOTING CHRISTIAN KNOWLEDGE,
NORTHUMBERLAND AVENUE, CHARING CROSS, W.C.;
43, QUEEN VICTORIA STREET, E.C.
BRIGHTON: 135, NORTH STREET.

NEW YORK: E. & J. B. YOUNG & CO.
1889.

CONTENTS.

CHAPTER		PAGE
I.	INTRODUCTORY	1
II.	CHALK-MAKERS, OR FORAMINIFERA	27
III.	LATTICE-WORKERS, OR POLYCYSTINA	66
IV.	SPONGE WEAVERS	109
V.	PLANT-ANIMALS, OR ZOOPHYTES	145
VI.	SEA-FAN MAKERS	180
VII.	CORAL BUILDERS	215
VIII.	CORAL REEFS, AND ISLANDS	247
IX.	SEA-MAT MAKERS	291
X.	TUBE-MASONS	314
XI.	EXCAVATORS	340

EXPLANATION OF PLATES.

PLATE I.—FORAMINIFERA.

Fig.
a. Rotalina Beccarii.
b. Rotalina turgida.
c. Rotalina concamerata.
d. Nonionina Jeffreysii.
e. Polystomella crispa.
f. Bulima pupoides.
g. Nodosaria radicula.
h. Nummulina planulata.
i. Textularia cuneiformis, var.
k. Globigerina bulloides.
l. Peneroplis planatus.
m. Cristellaria calcar.
n. Lagena vulgaris, var.
o. Cristellaria subarcuatula.
p. Miliolina seminulum, var. disciformis.
q. Rotalina inflata.
r. Textularia variabilis.
s. Polymorphina lactea.
t. Lagena vulgaris, var.
u. Biloculina ringens.
v. Miliolina bicornis, var. elegans.

Mostly after Williamson.

PLATE II.—POLYCYSTINS.

Fig.
a, b. Sethamphora ampulla.
c. Astractura aristotelis.
d. Trigonactura rhopalastrella.
e. Podocyrtis Schomburghii.
f Undetermined species.
g. Heliosestrum craspedotum.
h. Histiastrum gladiatum.
i. Acanthocyrtis mespilus.
k. Dictyophimus lucerna.

PLATE III.—SPONGE SPICULES.

Fig.
a. Euplectella × 90.
b. Hyalonema mirabilis × 83.
c. Undescribed sponge × 600.
d. Halichondria Ingallii × 260.
e. Spongilla cinerea × 600.
f. Spongilla cinerea.
g. Sponge unknown.
h. Sponge unknown.
i. Australian sponge.
k. Tethea robusta × 660.
l. Tethea Ingallii × 660.
m. Hyalonema mirabilis × 175.
n. Halichondria coccinea × 160.
o. Hymedesmia Johnstonii × 400.
p. Halichondria variantia × 1000.
q. Euplectella aspergillum × 90.
r. Hymeniacidon Bucklandii × 90.

Fig.
s. Ecionemia × 108.
t. Grantia × 180.
u. Halicnemia patera × 175.
v. Leuconea nivea × 660.
w. Halichondria infundibuliformis × 160.
x. Spongilla fluviatilis × 160.
y. Tethea muricata × 308.
    Mostly after Bowerbank.

PLATE IV.—GORGONIA SPICULES.
Fig.
a. Gorgonia species × 300.
b. Gorgonia papillifera × 300.
c. Gorgonia crinita × 300.
d. Gorgonia species × 300.
e. Plexaura racemosa × 300.
f, g. Gorgonia vatricosa × 100.
h. Gorgonia discolor × 100.
i, k. Gorgonia exserta × 100.
l. Rhipidigorgia flabellum × 200.
m. Pterogorgia setosa × 200.
n, o. Rhipidigorgia flabellum × 200.
p. Pterogorgia setosa × 200.
q. Hymenogorgia quercifolia × 300.
r. Pterogorgia suberosa × 200.
s. Xiphigorgia setacea × 200.
t. Leptogorgia viminea × 200.
u. Lophogorgia palma × 400.
v. Juncella juncea × 150.
w. Leptogorgia Boryana × 150.

Fig.
x. Juncella juncea × 150.
y. Juncella elongata × 150.
z. Pterogorgia betulina × 150.
aa. Paramuricea placomus × 50.
ab. Variety of same × 50.
ac. Muricea echinata × 50.
ad. Muricea lima.
ae. Eunicea plantaginea × 50.
af. Primnoa verticillaris × 50.
ag. Primnoa monilis × 50.
ah. Muricea fungifera × 100.
ai. Rhipidigorgia coarctata × 100.
ak. Plexaura porosa × 100.
al. Plexaurella dichotoma × 100.

Mostly after W. Savile Kent.

TOILERS IN THE SEA.

CHAPTER I.

INTRODUCTORY.

IT may as well be confessed at once that the first suggestion of this volume was derived from a perusal of the Rev. J. G. Wood's entertaining "Homes without Hands," strengthened by the observation that he had left the "Homes" included herein practically unnoticed, and that therefore the subject was still open for treatment. In some respects the animals are not so attractive, or interesting, as those with which the above-named volume is illustrated, but the "homes" are not less remarkable, and a brief summary of their structure and architecture, with a few details of the builders, may not be less acceptable to "lovers of nature."

The objection has been made, and may be repeated, that it is not an absolutely accurate designation to write of some of these as "homes" constructed by the animals for a residence, since they are, in many cases, the skeletons of the animals themselves, eliminated and deposited as the bones are built up in mammals. We are not disposed to insist strongly on the absolute meaning of terms, and are content to adopt a popular view, as most suitable to our purpose, and in a broad sense to accept these structures, in and about which the animals reside, as their homes, whether they are wholly located within them, or partially enclose them with their own flesh.

At first we had designed to confine ourselves absolutely to such animals as lived in communities, as for instance the coral-makers, the sea fanmakers, the animals which inhabit sponges, sea-mats, and those known at one time more particularly as zoophytes, but now more specially as Hydroid zoophytes, but by this arrangement we should have excluded those minute, but widely diffused, animals which construct those marvellous little shells known to every microscopist as Foraminifera and Polycystina. This would so much have diminished the value of our volume in the estimation of those for whom it was most of all designed, that the resolution was taken to widen the original scope so as to include all the minute marine animals which construct for themselves a permanent home. For whom, then, do we suppose that we are writing? it may be inquired, and for the information of those who are not leaving this chapter to be the last that is read, we may suggest that it is the large and increasing section of the nature-loving public who indulge in the use of the microscope, as a source of instruction and amusement, that awakens our sympathies and, as it seems to us, desire and require some introduction to the marvels of marine life, as a preliminary to more specific knowledge, the direction of which they will thereafter be better able to choose. Unless we are mistaken in the wants of this large group of workers and readers, they desire some volume which gives, within a reasonable compass, an outline of the structure and habits of a number of families of marine animals, not otherwise to be obtained except by wading through several volumes, and learning a copious vocabulary of technicalities. With this interpretation of what is desirable, we have attempted to fulfil the conditions in such a manner as to be able to appeal also to those who, without microscopical proclivities, seek an introduction to these "Toilers in the Sea."

Wherefore are coral reefs and their builders included, when they are tropical, and beyond the reach and experience of the ordinary British reader? Simply because our scope is wider than to provide a mere introduction to the marine zoology of our own shores, a fact accomplished beforehand, since such a restriction is hardly compatible with a general survey of ocean homes and their builders. And if it is conceded to be a legitimate project to endeavour to arouse a greater interest in the minute inhabitants of the ocean, which construct and leave behind them such marvellous structures, as evidence of their ingenuity and perseverance, it would have been most unwarrantable to have excluded one of the most interesting and remarkable groups. There are many charming books for the seaside, with which we had no ambition to compete; most delightful companions which any lounger at our numerous watering-places could not fail to appreciate and enjoy. The Rev. Charles Kingsley's "Glaucus, or Wonders of the Shore," is one of the most fascinating little volumes for young readers. Mr. P. H. Gosse is a veteran in the field of marine zoology, with half a dozen volumes; and we cannot forget Mr. G. H. Lewes's "Seaside Studies," besides a host of others, with more or less pretensions, and yet none of these appeared to touch, or more than touch, that phase of marine life to which these pages are devoted. And yet this can hardly be called a "seaside book" after the manner in which *they* are seaside books, although some knowledge of ordinary marine life would, perhaps, render it the more intelligible.

It is wholly unnecessary to advance any plea on behalf of the study of marine life, in the face of abundant evidence that every year increases its interest, and largely augments the number, not by any means a small one, of those who have made it their permanent choice. To those who have no desire for serious study, and no leisure withal, we would commend the perusal of the following passage which occurs in Quatrefages' charming volumes, and will merit a passing reflection :—" If you still preserve any of those illusions which, day by day are vanishing amid the turmoils of life, if you regret the dreams that have fled never to return, go to the ocean side, and there on its sonorous banks you will assuredly recall some of the golden fancies that shed their radiance over the hours of your youth. If your heart have been struck by any of those poignant griefs which darken a whole life, go to the borders of the sea, seek out some lonely beach, beyond reach of the exacting conventualities of society, and when your spirit is well-nigh broken with anguish, seek some elevated rock, where your eye may at once scan the heaving ocean and the firmament above ; listen to the grand harmonious voices of the winds and waves, as at one moment they seem to murmur gentle melodies, and at another swell in the thundering crash of their majesty ; mark the capricious undulations of the waves, as far as the bounds of the horizon, where they merge into the fantastic figures of the clouds and seem to rise before your eyes into the liquid sky above. Give yourself up to the sense of infinitude which is stealing over your mind, and soon the tears you shed will have lost their bitterness; you will feel ere long that there is nothing in this world which can so thoroughly alleviate the sorrows of the heart as the contemplation of nature, and of the sublime spectacle of creation, which leads us back to God."[1]

Thousands of those who rush every year to the sea shore, embued with no particular feeling but that of enjoying themselves, in a manner of their own selection, return with a lively sense of physical benefit, and we would hope, in many cases, of intellectual also. The latter class might be considerably augmented, without any diminution in the results attributed to the former, provided they could be stimulated to the use of the observing eye. This could be hardly accomplished better than by indicating the direction in which the observing eye should be turned, and if this be achieved this volume will justify itself.

No event of modern times has so greatly stimulated the student of marine life as the publication of the results of the "Challenger" expedition, and to some extent this has reacted upon those who cannot exactly be described as students, but only as lovers

---

[1] Quatrefages, "Rambles of a Naturalist," vol. i. p. 120.

of nature. There was a period, not very remote, when the botany and zoology of the shore was held to represent the whole of marine life. Then low water-mark was almost the limit of investigation, and for the majority the absolute limit. Sea-weeds, anemones, shells of molluscs, zoophytes, and stranded sponges, were, apart from fish and crustacea, the totality of marine life. But now these are almost discarded in favour of the denizens of the depths of the sea. The waves of enthusiasm at different times set in different directions, determined perhaps by some single circumstance, and in that direction which they assume, are probably persistent for a considerable period, until some new impulse is given, and the current is changed. To know something of the most extraordinary of the "Toilers in the Sea," the most expert builders of "ocean homes," we must go beyond the littoral zone, and explore the dark abysses, where the light of the sun and the eye of man has never penetrated.

All the preliminary facts which it is necessary to call to remembrance here, in reference to the ocean in which our organisms flourish, may be indicated by a short retrospect, for the most part abstracted from Sir Wyville Thomson's "Depths of the Sea." Taking for granted that three-fourths of the surface of the globe is covered by the sea, and until very recent times so little was known of its depths, and so much imagined, it may be well to ascertain what data modern investigation and discovery have given us, and how far old notions have been superseded. Popular opinion originally favoured the supposition that beyond a certain depth there was only "a waste of utter darkness, subject to such stupendous pressure as to make life of any kind impossible, and to throw insuperable difficulties in the way of any attempt at investigation." When Carpenter and Thomson in 1868 found that they could work "not with so much ease, but with as much certainty, at a depth of 600 fathoms as at 100, "great hopes were entertained in the results, which were further strengthened when in 1869 they carried their "operations down to 2,435 fathoms (14,610 feet), nearly three statute miles, with perfect success." Thus was the operation of deep-sea investigation demonstrated as possible, and the celebrated "Challenger" expedition, soon to follow, translated it into fact. With the preliminary experience of the "Porcupine" expedition to guide him, Sir Wyville Thomson wrote :—" For the bed of the deep sea, the one hundred and forty millions of square miles which we have now added to the legitimate field of Natural History research, is not a barren waste. It is inhabited by a fauna more rich and varied on account of the enormous extent of the area, and with the organisms in many cases apparently even more elaborately and delicately formed, and more exquisitely beautiful in their soft shades of colouring, and in the rainbow tints of their wonderful phosphorescence, than the fauna of the well-known belt of shallow water teeming with innumerable invertebrate forms which fringes the land. And the forms of these hitherto unknown living beings, and their mode of life, and their relations to other organisms, whether living or extinct, and the phenomena and laws of their geographical distribution must be worked out."[1]

The most important investigations of the sea depths were those by John Ross, in Baffin's Bay, in 1818; by James Ross, in the Pacific Ocean, in 1843; by Lieut. Joseph Dayman, in the North Atlantic, in 1857; by Dr. Wallich, in the North Atlantic, in 1860; by Chydenius and Torell, near Spitzbergen, in 1861; by Carpenter, Jeffreys, and Thomson, in the North-East Atlantic, in 1868 and 1869; by Pourtales, in the Gulf Stream off Florida, in 1869; and the famous "Challenger" Expedition, from 1873 to 1876. Between the first and last of these, extending over three-quarters of a century, the steps were gradual, but all tending to a distrust of the notion that there was a zero of animal life at a limited

---

[1] "The Depths of the Sea." By C. Wyville Thomson. London: Macmillan & Co. (1873), p. 4.

depth. The earliest record of living animals from any depth approaching 1,000 fathoms, was that of John Ross, in Baffin's Bay, in 1818, of which his official account states :—" Soundings were obtained correctly in 1,000 fathoms, consisting of soft mud, in which were worms." Later on (1843) Sir James Ross reports :—" I have no doubt that, from however great a depth we may be enabled to bring up the mud and stones of the bed of the ocean, we shall find them teeming with animal life ; the extreme pressure at the greatest depth does not appear to affect these creatures ; hitherto, we have not been able to determine this point beyond a thousand fathoms, but from that depth several shell-fish have been brought up with the mud." In 1860, Dr. Wallich records that thirteen star-fishes came up from a sounding of 1,260 fathoms. In his "Diary," published in 1862, he advocated the view that the conditions of the bottom of the sea were not such as to preclude the possibility of the existence of even the higher forms of animal life. Passing to the "Porcupine" Expedition, from 1868 to 1870, we have the record that "Sixteen hauls of the dredge were taken at depths beyond 1,000 fathoms, and in all cases life was abundant. In 1869, we took two casts in depths greater than 2,000 fathoms. In both these life was abundant ; and with the deepest cast, 2,435 fathoms, off the mouth of the Bay of Biscay, we took living, well marked, and characteristic examples of all the five invertebrate sub-kingdoms. And thus the question of the existence of abundant animal life at the bottom of the sea has been finally settled, and for all depths, for there is no reason to suppose that the depth anywhere exceeds between three and four thousand fathoms; and if there be nothing in the conditions of a depth of 2,500 fathoms to prevent the full development of a varied fauna, it is impossible to suppose that even an additional thousand fathoms would make any great difference."

"The conditions which might be expected principally to affect animal life at great depths of the sea are pressure, temperature, and the absence of light, which, apparently, involves the absence of vegetable food. Beyond the zone surrounding the land, speaking generally, the average depth of the sea is 2,000 fathoms, or about two miles; as far below the surface as the average height of the Swiss Alps. In some places the depth seems to be considerably greater, possibly here and there nearly double that amount; but these abysses are certainly very local, and their existence is even uncertain, and a vast portion of the area does not reach a depth of 1,500 fathoms."

Having given these general estimates of the depth of the ocean, Professor Thomson deals with the three supposed conditions inimical to animal life. And first, as to pressure. "There was," he says, "a curious popular notion, in which I well remember sharing when a boy, that, in going down, the sea-water became gradually, under the pressure, heavier and heavier, and that all the loose things in the sea floated at different levels, according to their specific weight, forming a kind of false bottom to the ocean, beneath which there lay all the depth of clear, still water, which was heavier than molten gold."

"The conditions of pressure are certainly very extraordinary. At 2,000 fathoms, a man would bear upon his body a weight equal to twenty locomotive engines, each with a long goods train loaded with pig-iron. We are apt to forget, however, that water is almost incompressible, and that, therefore, the density of sea-water at a depth of 2,000 fathoms is scarcely appreciably increased. But an organism supported through all its tissues on all sides, within and without, by incompressible fluids at the same pressure, would not necessarily be incommoded by it. We sometimes find when we get up in the morning, by a rise of an inch in the barometer, that nearly half a ton has been quietly placed upon us during the night, but we experience no inconvenience, rather a feeling of exhilaration and buoyancy, since it requires a little less exertion to move our bodies in the denser medium. At all events, it is a fact that the animals of all the invertebrate classes, which abound at a depth of 2,000 fathoms, do bear that extreme pressure, and that they do not seem to be affected by it in any way."

Secondly, as to temperature. It was generally understood, until about twenty-five years ago, that whatever the variation of surface temperature might be, there was a certain depth at which the temperature was permanent at 4° C., which is the temperature of the greatest density of fresh water. On this point it is unnecessary to recapitulate the details of a long chapter, but to accept the broad conclusions at which Professor Thomson arrived. That, "instead of there being a permanent deep layer of water at 4° C., the average temperature of the bottom of the deep sea, in temperate and tropical regions, is about 0° C., the freezing point of fresh water; and that there is a general surface movement of warm water, produced probably by a combination of various causes, from the equatorial regions towards the poles, and a slow under-current, or rather indraught of cold water from the poles towards the equator." "The temperature of the sea apparently never sinks, at any depth, below—3°·5 C. (3½° Cent. below the freezing point of fresh water), a degree of cold which, singularly enough, is not inconsistent with abundant and vigorous animal life, so that in the ocean, except perhaps within the eternal ice-barrier of the antarctic pole, life seems nowhere to be limited by cold."

"Edward Forbes pointed out long ago the kind of inverted analogy which exists between the distribution of land animals and plants, and that of the fauna and flora of the sea. In the case of the land, while at the level of the sea there is, in temperate and tropical regions, a luxuriant vegetation with a correspondingly numerous fauna, as we ascend the slope of a mountain range the conditions gradually become more severe; species after species belonging to the more fortunate plains beneath disappear, and are replaced by others whose representatives are only to be found on other mountain ridges, or on the shores of an arctic sea. In the ocean, on the other hand, there is along the shore line, and within the first few fathoms, a rich and varied flora and fauna, which participates and sympathises in all the circumstances of climate which affect the inhabitants of the land. As we descend, the conditions gradually become more rigorous, the temperature falls, and alterations of temperature are less felt. The fauna becomes more uniform over a larger area, and is manifestly one of which the shallower water fauna of some colder region is to a great extent a lateral extension. Going still deeper, the severity of the cold increases, until we reach the vast undulating plains and valleys at the bottom of the sea, with their fauna partly peculiar and partly polar—a region the extension of whose extreme thermal conditions only approaches the surface within the arctic and antarctic circles."

Finally, as to the absence of light. Very little exact knowledge on this point has been obtained. From recent experiments "it would appear that the rays capable of affecting a delicate photographic film are very rapidly cut off, their effect being imperceptible at the depth of only a few fathoms. It is probable that some portions of the sun's light possessing certain properties may penetrate to a much greater distance, but it must be remembered that even the clearest sea-water is more or less tinted by suspended opaque particles, and floating organisms, so that the light has more than a pure saline solution to contend with. At all events it is certain that beyond the first 50 fathoms, plants are barely represented, and after 200 fathoms they are entirely absent. There seems to be little or no light at the bottom of the sea, and there are certainly no plants except such as may sink from the surface, but the bottom of the sea is a mass of animal life."[1]

It has been seen that pressure presents no obstacle to animal life at great depths, that temperature is not carried below that at which animal life may be sustained, but that there is practically an absence of light, and consequently of vegetable life.

---

[1] "The Depths of the Sea." By C. Wyville Thomson. Introductory Chapter, pp. 1–48.

Examinations of the deep-sea bottom, made at intervals during the past eighty years, have revealed the fact that the ocean bottom at great depths, say from 500 to 3,000 fathoms, consists largely of a tenacious mud, in which a great number of animals exist, obtain nourishment, and increase their species. Various hypotheses have been offered as to the manner in which such conditions have been made possible. How nourishment could be furnished at such extreme depths has been the initial question. Dr. Wallich suggests that the Rhizopods of the deep sea must have the faculty of separating the elementary constituents of their bodies from the surrounding medium.[1] It has been contended that only organisms containing chlorophyll "have the power of producing albuminoid compounds from carbonic acid, water, ammonia, and nitric acid." Dr. Carpenter was inclined to accept the view propounded by Thomson, that "the Protozoa of the deep sea are nourished by protoplasm which is diffused through the whole mass of sea-water, renewed constantly by the plants and animals living at its surface, and penetrating by diffusion to its greatest depths."[2] Gwyn Jeffreys suggested that the decomposed organic mass was derived from animals

---

[1] "North Atlantic Sea Bed" (1862) p. 130.
[2] *Nature*, March 31, 1870.

which have sunk down from the surface.[1] Lieutenant Maury came to a similar conclusion, as would appear from his remarks that "the ocean swarms with living creatures, especially between and near the tropics. The remains of their myriads are carried on and collected by the currents, and in the course of time deposited like snowflakes on the bottom of the sea. This process, going on for centuries, has covered the depths of the ocean with a mantle of organisms as delicate as hoar-frost, and as light as down in the air."[2]

Professor Karl Mobius has detailed a somewhat similar view, that "the plants which have grown in the higher slopes sink to the bottom after they have died, gradually break up into smaller and smaller portions, and finally glide down into the greatest depth that they can attain. This organic, and chiefly vegetable, mass, is what renders the mud-region inhabitable by a great number of animals,— in the first place, by those which feed upon decaying matters, and then for others which devour the dirt-eaters. In this way we find it easy to explain the quantities of individuals (at the first glance quite astonishing) which may be got out of the mud of the greater depths; for the mass which serves them

---

[1] *Nature*, Dec. 9, 1869.
[2] "Physical Geography of the Sea" (1869), § 617.

as a dwelling-place, at the same time contains an enormous store of nourishment for them. The same thing must take place in all seas. In the shallower regions which immediately surround continents and islands, great masses of Algæ grow wherever there are rocks and stones. In the warmer seas there is an enormous floating Sargasso-life. Only a small portion of these plants is directly eaten by animals, or thrown upon the shore. Most of them die where they have lived, or, after they have been carried away by currents and winds, lose the gases which make them lighter than sea-water, sink down, and become finally decomposed into a soft mass. With the sinking organic materials are, of course, intermixed the remains of Testacea, and the fine inorganic soil constituents of the higher regions, which the currents of flood and ebb and the waves are unceasingly triturating. This muddy mixture must move down towards the deeps upon the sloping sea bottom in the neighbourhood of the coasts, from purely mechanical causes, until the weight and mutual adhesion of the individual particles present so much resistance to the pressure of the masses following them from above that equilibrium is produced." And again, "dead plants, fragments of shells, and sand, are heaped one upon the other to a height of feet or fathoms. The alternation of flood and ebb, and the winds, keep the upper strata of the water in constant movement, and produce oscillations up and down, even in the lower ones, by increasing or diminishing the column of water resting on the bottom. The differences of temperature which are dependent on the alternation of day and night, on changes of weather, and the course of the seasons, cause expansions and displacements of the constituents of the bottom. Into the greater depths, where these forces can operate but rarely or slightly, or even not at all, the currents of sinking water, which has become heavier than the subjacent strata, by cooling or increase of its amount of salt, penetrate." Finally, he says, "Of all the movements which convey organic materials to the sea-bottom, descending currents are evidently among the most efficacious. Their operation falls precisely in the most suitable season for this purpose; it commences after the annual development of the marine vegetation in the temperate and cold zones has attained its maximum, when strong and long-continued storms gather their chief harvest in the fields of *Zostera* (grass wrack) and tangle, and the bottom of the sea is disquieted to a greater depth than usual."[1]

This is, in effect, perhaps the most reasonable and

---

[1] "Whence comes the Nourishment for the Animals of the Deep Seas," by Prof. Karl Mobius, in "Annals Nat. Hist.," vol. viii. (1871), p. 193.

feasible hypothesis yet advanced to account for the existence, by furnishing nourishment, to the vast bulk of minute animal life which has been found to exist in the greatest depths yet reached in the bottom of the "deep, deep sea."

Remote as may be its connexion with all that follows, still its association with the mysterious depths of the ocean may justify a slight reference here to the controversy concerning the equally mysterious *Bathybius*, which, but a short time since, was in the mouth of all who spoke or thought of the bottom of the sea. It was in 1868 that a distinguished naturalist announced the supposed discovery of a new organism, at great depths in the Atlantic, to which he gave the name of *Bathybius*.[1] Shortly afterwards it was again referred to in a "Preliminary Report" in these terms:—"The examination which Professor Huxley has been good enough to make of the peculiarly viscid mud, brought up in our last dredging at the depth of 650 fathoms, has afforded him a remarkable confirmation of the conclusion he announced that the coccoliths and coccospheres are imbedded in a living expanse of protoplasmic substance," and, further, "that there seems adequate reason for regarding this *Bathybius*

---

[1] "On Some Organisms living at Great Depths in the North Atlantic," by Professor Huxley, in *Quart. Journ. Micr. Sci.*, vol. viii. (1868), p. 203.

as one of the chief instruments whereby the solid material of the calcareous mud which it pervades is separated from its solution in the ocean-waters." Again, in 1870, Professor Huxley emphasised the discovery by remarking that "the *Bathybius* formed a living scum or film on the sea-bed extending over thousands upon thousands of square miles; evidence of its existence having been found throughout the whole North and South Atlantic, and wherever the Indian Ocean had been surveyed, so that it probably forms one continuous scum of living matter girding the whole surface of the earth."

In 1875 a modification of these views seemed to be imperative, and Professor Huxley proceeded to explain:[1] — "Professor Wyville Thomson informs me that the best efforts of the 'Challenger's' staff have failed to discover *Bathybius* in a fresh state, and that it is seriously suspected that the thing to which I gave that name is little more than sulphate of lime, precipitated in a flocculent state from the sea-water, by the strong alcohol in which the specimens of the deep-sea soundings which I examined were preserved. The strange thing is that this inorganic precipitate is scarcely to be distinguished from precipitated albumen, and it resembles, perhaps even more closely, the proliferous pellicle on the

---

[1] *Nature*, August 19, 1875.

surface of a putrescent infusion (except in the absence of all moving particles) colouring irregularly, but very fully, with carmine, running into patches with defined edges, and in every way comporting itself like an organic thing. Professor Thomson speaks very guardedly, and does not consider the fate of *Bathybius* to be as yet absolutely decided. But since I am mainly responsible for the mistake, if it be one, of introducing this singular substance into the list of living things, I think I err on the right side in attaching even greater weight than he does to the view which he suggests."

This, then, would seem to be an end of the matter, and it may be presumed that *Bathybius*, as an organised substance, was altogether a mistake, as confirmed by the subsequent strictures by Dr. G. C. Wallich,[1] which, apart from the little personal animus that leaven them, may be taken as conclusive. Some concluding remarks have such an intimate relationship with the study of marine life, and contain salutary hints worthy of remembrance in this connexion, that no apology need be made for quoting them. Dr. Wallich says :—" I fully appreciate the extreme difficulties under which he (Huxley) worked when analysing material unquestionably altered in its most

---

[1] "On the True Nature of the so-called Bathybius," by Dr. G. C. Wallich, in "Annals Nat. Hist.," vol. xvi. (1875), p. 322.

important characters by the admixture of alcoholic preservative solutions. I can attest, from personal and long-continued experience, that it is simply impossible to arrive at a correct knowledge of the characters of the recent unadulterated material from material that has been thus preserved. The fact is that there is as marked a distinction between the aspect of pure fresh sponge-protoplasm, for example, seen instantly on its arrival at the surface, and its aspect a very brief period afterwards, as there is between living Foraminifera and Polycystina of the open ocean immediately after capture, and after they have been consigned to some preservative solution. In addition to other important changes produced in the protoplasm of the Protozoa, both marine and freshwater, by being long kept or preserved in such preservative solutions as alcohol, when calcareous matter exists in solution, molecular changes take place, the normally homogeneous protoplasm then frequently being converted into minute globular masses, which, when seen under the microscope resemble sago-grains in miniature, and may readily be mistaken for molecular granules of the organism within or upon which they occur. I can produce specimens of Polycystina, and, to a certain extent, of Foraminifera, the rich and varied brilliancy of colour in which has been retained for years, in some cases, even when mounted in balsam ; but there all identity in the appearance of the soft parts ends; and so it must be with any protoplasmic matter. Let this episode of *Bathybius* be remembered as a caution by future investigators, and it will have served a good purpose."

It would avail nothing to enter upon the discussions which were at one time, not long distant, so general as to the basis of life. After all that has been said, or can be said, the mystery of life is a mystery still. On the one hand were those who maintained that the formation of living beings out of inanimate matter, by the conversion of physical and chemical into vital modes of force, was going on daily and hourly; that, in fact, the actions of living beings are not due to any mysterious vitality, or vital force or power, but are physical and chemical in their nature. On the other hand are those who contend that such an explanation is illusory and insufficient. One of these writes thus :—" In all living beings there exists matter in a peculiar state which we call *living*. This living matter manifests phenomena which are different from any phenomena proved to be due to the operation of any known laws. It moves in a manner which cannot be explained by physics. Changes are effected in its composition which cannot be accounted for, and various substances are formed by it which may exhibit structure, properties, and a capacity for acting in a manner which is peculiar to living beings, and cannot be imitated artificially, or satisfactorily explained. It takes up non-living matter in solution and communicates its wonderful properties to it. Having increased to a certain size, the mass of living matter divides into smaller portions, every one of which possesses the same properties as the parent mass, and in equal degree."

"Scientific investigators have hitherto failed to discover any laws by which these facts may be accounted for. But rather than ignore or misrepresent them, or affirm anything concerning them which we cannot prove, as some have done, it seems to me preferable to resort provisionally to hypothesis. In order to account for the facts, I conceive that some directing agency, of a kind peculiar to the living world, exists in association with every particle of living matter, which, in some hitherto unexplained manner, affects temporarily its elements, and determines the precise changes which are to take place when the living matter again comes under the influence of certain external conditions." In higher animals, besides giving rise to the phenomena above referred to, every instant during life in every part of the organism, this supposed agency or power, acting under certain circumstances at an early period of development, so disposes the material which it governs, that mechanisms result of most wonderful structure, admirably adapted, as they have been evidently actually designed, for the fulfilment of definite purposes. These mechanisms were anticipated, as it were, from the earliest period, and their formation provided for by the preparatory changes through which the structures had to pass before perfect development could be attained. Can these phenomena be accounted for except through the influence of some wonderful power or agency such as we are now contemplating." [1]

However interested we may feel in the solution of the problem of life, there is little prospect of gathering much to assist in that solution from the succeeding chapters, which had no such design in their composition. The foregoing allusion to the subject is almost parenthetical, induced by the mention of the so-called *Bathybius*. It may now be abandoned, leaving us free to pursue, as far as we can within the prescribed limits, the history of the little animals which we have brought together under the designation of "Toilers in the Sea."

---

[1] "Protoplasm, or Life, Force, and Matter," by Lionel S. Beale. London (1870), p. 89.

CHAPTER II.

CHALK MAKERS, OR FORAMINIFERA.

ANIMALS and plants are classified, for scientific purposes, in a regular series of groups, according to their structure and relationship, either in an ascending scale, from the simplest to the most complex, or descending, from the highest or most complete, to the lowest or simplest forms. This classification undergoes modification from time to time, so as to be in harmony with the progress of knowledge, seeking to represent, as nearly as possible, a graduated sequence of animal or vegetable life, upwards or downwards, as the case may be, beginning or ending with the simplest forms. It is of little moment whether the series be an ascending or descending one, so long as the sequence is in accord with ascertained facts. Whether from acquired habits of order, or from prejudice, or from conviction of its many advantages, the biologist is always apt to think and write with this classification, almost unconsciously, before him; ánd it seems to him the most natural thing in the world to proceed step by step, up or down, the orthodox ladder. It is in deference to this feeling that we commence with the simplest or *lowest*, if that term be preferable, of animals that construct for themselves " homes in the sea."

Those who have made acquaintance with the simple forms of animal life to be found in fresh waters, are cognizant of the existence of a strange little organism to which the name of *Amœba* or Proteus has been given. They are common enough in all water containing decaying vegetable matter, and are generally regarded as one of the simplest forms of animal life to be found in our " ponds and ditches." The general appearance is that of a microscopical particle of jelly, transparent amorphous jelly, or, as Dr. Carpenter has described it, " a little particle of homogeneous jelly arranging itself into a greater variety of forms than the fabled Proteus, laying hold of its food without members, swallowing it without a mouth, digesting it without a stomach, appropriating its nutritious material without absorbent vessels or a circulating system, moving from place to place without muscles feeling (if it has any power to do so) without nerves propagating itself without genital apparatus, and not only this, but in many instances forming shelly coverings of a symmetry and complexity not surpassed by those of any testaceous animal." The latter part of this paragraph does not refer to the true

Amœba, but to the allied forms to which our remarks will shortly be confined. Suffice it to say that Dr. Carpenter recognises the Amœban form of simple animal as that of the Foraminifera which secrete a calcareous skeleton or shell. By commencing with a reference to this well-known organism (fig. 1), although it may in some points differ from the shell-making species, an acquaint-

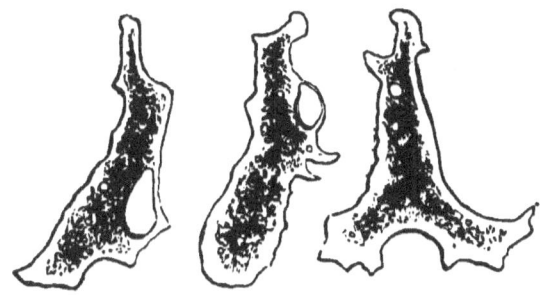

FIG. 1.—PROTEUS, or AMŒBA.

ance with the latter may be facilitated, and through the Amœba we may obtain an introduction to the Foraminifera and Polycystina, which are of the same kindred. It is of little consequence to us that some have regarded the Amœba as lower in the scale of creation than the shell-bearers, whilst others consider it superior on account of "something akin to organisation having reached its highest limit."[1] It is admitted, on all hands

---

[1] Dr. Wallich, in *Monthly Microscopical Journal*, vol. i. (1869), p. 231.

that the Amœba is a minute jelly-like animal, which has no special form of its own, but is subject to incessant changes of form, by thrusting out portions of its gelatinous substance, and retracting others, moving slowly from place to place by means of these foot-like extensions of the body, called *pseudopodia*, or false feet. "On coming in contact with any particle of organic substance, on which the Amœba is inclined to feed, and they obviously have some choice in this respect, the sarcode proceeds to surround it, so that it is soon enclosed, along with some water, within the body. Most observers agree in believing that any part of the body may thus incept the food, though some have believed in an oral aperture (mouth). No such aperture has been discovered, and the evidence does not indicate its existence. It also appears indisputable that when the Amœba has extracted the nourishment from the food, it ejects what remains, but not at the spot where the food entered. This is done at the posterior extremity of the body, where also the contractile vesicles discharge themselves. Writers speak of an anal outlet at this point, but there is no reason for believing in its presence. The particles appear to be simply forced out through the sarcode."[1]

[1] "The Amœba," by Prof. W. C. Williamson, in *Popular Science Review*, vol. v. (1866), p. 188.

It has been contended that there exists an internal circulation in the body, because when the animal pushes out a lobe, or prolongation, a rush of granules will be seen flowing in the direction of the new foot or lobe. Dr. Wallich contends that the flow is merely mechanical, and this view is generally accepted. "In the large pseudopodia of *Amœba*," he says, "we see at once that the appearance of an advancing and a retrograde current is due to the fact that the lower surface of the organism is fixed, as it were, to the plane on which it rests, whereas its upper or free surface only is being projected into pseudopodial extensions. In short, the effect is identical in character with that which would present itself, if, after filling a transparent and highly-elastic bladder, with some viscid and transparent fluid in which granular particles of slightly greater specific gravity than the fluid itself were suspended, we were to roll it slowly along a flat surface. In such a case, the upper stratum of granules would in reality be moving forwards in the direction in which the bladder was being rolled; whereas the inferior stratum, although at rest, would appear to be retrograding; for the same reason that when a railway train is slowly and steadily put in motion, the platform appears to be moving from the observer seated in one of the carriages. Now, as regards the cyclosis, this result could not take place, if the two phenomena—namely, the vital contractility of the protoplasm itself, and the circulating force by means of which the granules are impelled,—acted independently one of the other. Did they act independently, any cessation or alteration in the one would not necessarily involve a cessation or alteration in the other, but the circulation of the granules would continue unchecked even when the protoplasmic mass had obtained a state of perfect rest. And, notably, when the direction in which the protoplasmic mass had for a time been moving became suddenly reversed, the direction of the granular movement would remain unaltered, at least for a period, were the force producing it an independent one. But the direction which the granules continue to take under these circumstances becomes immediately reversed also, proving thereby that it simply follows the direction imparted to it by the protoplasm. It only remains to be stated that these phenomena are observable whenever a fresh pseudopodium is projected; every modification in the direction taken by the current of granules being distinctly referable to some corresponding change in the form being assumed by the protoplasmic body generally.[1]

---

[1] Dr. G. C. Wallich, "On the Rhizopoda," in *Monthly Micro. Journ.*, vol. i. (1869) p. 233.

Without staying here to pursue further inquiries as to the structure and habits of the naked *Amœbæ*, we may pass to the simplest condition of an Amœba-like body enclosed within a shell, or house, which it constructs for itself by accumulating grains of sand and other foreign particles which adhere to the membraneous covering of the animal. The species of *Difflugia* are also found in fresh water, and are rather common, but not being considered generally attractive, are but little studied (fig. 2). The coat, or "test," is usually ovoid, or flask-shaped, with a notched aperture at the end, its foundation is a thin transparent colourless membrane, and to this the particles adhere to form the encrusted house in which the animal dwells; the pseudopodia are protruded from the orifice only.

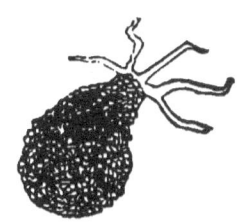

FIG. 2.—DIFFLUGIA.

The transition is a short and easy one to the Foraminifera, those minute marine animals of very near kindred to *Amœba*, who construct for themselves a calcareous shell, or skeleton, many of which in form resemble very microscopical snail shells. (Plate I.) From the perforations which the walls of these shells present they were called *Foraminifera*, or pore-bearers and though small, they are, or have been, probably the most numerous of all created things, and have performed no insignificant part in the history of the world. For the present it will be sufficient to note that they are brought up from the deep-sea bottom everywhere, and that myriads of them are still living in the ocean, even as myriads have flourished in the seas of indefinable past ages. Many of these elegant little shells are individually not to be distinguished by the naked eye; but they partake of a variety of forms, some consisting of a single chamber, and others of a great many, but in all, their composition, like that of the oyster and cockle, is principally carbonate of lime, such as we know it in chalk and limestone, of which these shells are important factors.

Some of the shells consist of a single chamber, others of several, and still others of a very considerable number. The many-chambered shells are at first simple, or, as it seems probable with the rudiments of two or three others, the greater part of the chambers being added consecutively by a kind of budding, the primary chamber being the smallest. These chambers are not absolutely isolated, but the walls, or divisions, are perforated, and through these perforations there is a "continuity of protoplasm," a union of all the segments of the compound body which inhabits them; so that in such species as do not possess perforations in the external walls of the shell, through which to extend their pseudopodia, nutrition may be supplied down to the furthest chamber, by means of the minute threads of communication, from the open mouth of the shell, and the pseudopodia operating therefrom.

We may regard the animal as of close kindred to, although not absolutely the same, as an *Amœba*, from which it differs in its pseudopodia, or voluntary extensions of the body, being thin delicate threads, sometimes so thin as to be scarcely visible, and *not* obtuse lobes; and also in the absence of a nucleus. For all practical purposes we may assume that it is somewhat of an *Amœba*, enclosed in a shell. And here again we must apply a reservation, since at certain times, if not always, the substance (or sarcode) of the animal flows over and envelopes the whole external surface of the shell, and hence, as Dr. Carpenter observes, " it is by no means impossible that the digestive process may really be performed in this external layer, so that only the products of digestion may have to pass into the portion of the sarcode body occupying the cavity of the shell. A recognition of this fact has led some writers to protest against regarding the shell as a home constructed by the animal to inhabit, since, as they affirm, it is a skeleton built up within the body of the same materials, and in the same manner, as the bónes of the skeleton in a mammal.

" The sarcode body of such Foraminifera as have

been observed in the living state,"[1] Dr. Carpenter says, "is more or less deeply coloured; its tint being in some instances a yellowish brown, in other cases a crimson red. This colour seems in some instances to be uniformly diffused through the whole mass of the sarcode, probably owing to the fine state of division of the particles which possess it; but in the larger forms it occurs in much larger and more scattered masses, which appear sometimes to be collections of granules, and in other cases to be vacuoles filled with a coloured liquid. In the Foraminifera, with many-chambered shells, it is nearly always to be observed that the colour is deepest in the segments of the body which occupy the oldest chambers, and that it fades progressively in the segments which intervene between these and the one that occupies the last-formed chamber, which is often nearly colourless. There is strong reason to believe that the colouring material is directly derived from external sources, though modified in some cases by the agency of the animal itself."[2]

---

[1] The soft substance of which the body of these animals is composed was called by Dujardin *sarcode*, or rudimentary flesh, but some authors prefer to call it *protoplasm*, because, as Dr. Carpenter observes, "it is nothing else than *protoplasm*, in which every form of animal structure has its origin, and from which it is evolved by a process of gradual differentiation."

[2] "Introduction to the Foraminifera," by Dr. W. B. Carpenter (1862), p. 32.

The functions of the pseudopodia, or filamentous prolongations of the sarcode body (fig. 3) would appear to be, in the first instance, to obtain nourishment for the animal, and in the next place to attach themselves to any given spot, or to serve as means of locomotion from place to place. Whether they are organs of feeling, or that the animals are in possession of the sense of feeling, or, if so, to what extent, must be left an open question.

Some naturalists have doubted the power of

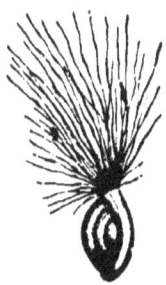 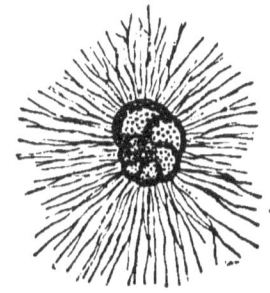

FIG. 3.—DISCORBINA WITH PSEUDOPODIA. FIG. 4.—MILIOLA WITH PSEUDOPODIA.

Foraminifera to move themselves from place to place by means of their pseudopodia (fig. 4). On this point the positive observations by so experienced an observer as Mr. P. H. Gosse may be accepted as conclusive. He obtained at Weymouth entangled amongst tufts of minute seaweeds a number of specimens of *Polystomella*. "These," he says, "were always found, a few hours after the weed had been deposited in my vases, adhering to the glass, with the pseudopodia extended in opposite directions. Very frequently these tiny atoms were found, in the morning, two, three, or even four inches up the sides of the perpendicular glass vases, having crawled this distance in the course of the night. And they never remained long stationary; the next morning would find tnem in some remote part of the glass. The night was manifestly their time of activity." And again, 'One of my tanks was literally swarming with a species of *Polymorphina*. The individuals were of various dimensions; a large number having attained the adult size. They studded the sides of the vessel, the stones, and the slender weeds, adhering to the filaments of the latter in such profusion as to cover the whole contents of the vessel with white dots, conspicuous even upon the most cursory glance. These, like the others, were constantly roaming; they crawled up and down the stems of the algæ, and over the various objects in the tank, never remaining long in one station. On removing one from the vessel to an aquatic cell for microscopic examination, it was found to be entirely withdrawn; but, in the course of a few minutes the pseudopodia were seen to be protruding their tips, and then they gradually stretched and expanded their lines and films of delicate sarcode, till in the course of a few hours these would sometimes reach almost from side to side of the glass cell. The extension was principally in two opposite directions, though the branched and variously connected films often diverged considerably, giving to the whole a more or less fan-like figure."[1]

The whole method of reproduction requires further investigation, since little more is known than was announced in 1856, and further confirmed in 1861, of the production of living young by two or three species. "Remarking that an individual had become stationary for several days, and enveloped, as is not unusual, in a thin layer of brownish slime, Max Schultze paid particular attention to it. At the end of a few days, after it had become quiescent, he noticed that minute spherical, sharply-defined granules were detached from the brownish slimy envelope, and in the course of a few hours the animal was surrounded by about forty of these corpuscles, which gradually became more and more widely separated from it. Microscopic examination of these bodies proved that they were young. When viewed by transmitted light, they presented a pale yellowish-brown calcareous shell, consisting of a central globular portion, partly surrounded by a closely-applied tubular part, and having no septum in the interior. In a short time the young animals protruded their

---

[1] "On Locomotion in the Foraminifera," by P. H. Gosse, in "Annals Nat. Hist.," vol. xx. (1857) p. 365.

contractile processes from the opening of the shell, and crawled about upon the object-glass. The parts of the body enclosed in the transparent shell could be examined with great accuracy under the highest magnifying powers, and were seen to consist of a transparent, very finely-granular, colourless material, of which the protruded filaments were an immediate continuation, and in which were imbedded minute sharply-defined granules, protein and fat molecules, some of considerable size, and angular, like the vitelline plates in the ovum of fishes. From the circumstances under which these young (*Miliolidæ*) made their appearance, it might be concluded that they must necessarily quit the parent in a tolerably perfect condition, and that it was probable they acquire the calcareous shell whilst still within the mother." The shell of the parent was broken up and only found to contain a little granular matter. " The almost complete absence of any organic contents in the shell of an individual, which from eight to fourteen days previously was creeping about, renders it probable that the whole, or, at any rate, the main part of the body, had been transformed into young ones."[1]

Subsequently, the same author recorded similar experiences in other species observed in a bottle of

---

[1] "Observations on Reproduction of the Rhizopoda," by Prof. Max Schultze. Abstracted in *Quart. Journ. Micr. Sci.*, vol. v. (1857) p. 220.

sea-water. Noticing suspicious circumstances in some of the individuals, he cleaned and examined one, which he found to contain coarsely granular yellowish-brown substance, the true nature of which could not be ascertained, on account of the opacity of the shell. He then proceeded to break up the shell, in which he counted ten chambers, and was not a little astonished, after detaching the first fragments of the shell, when some small three-chambered Foraminifera made their appearance, of which, after crushing and breaking up the mother, no less than from twenty to thirty were brought to light. They were all of nearly equal size, and each shell was three-chambered, the earliest chamber containing brownish-yellow substance, the others colourless. The shell appeared very thin and brittle, but immature. The artificial birth was too early for the young animals, and they manifested no signs of life. He then proceeded to watch the remaining specimens in the bottle, and soon observed in the neighbourhood of two of them an accumulation of small granules, which, when detached and placed under the microscope, were found to be also small three-chambered Foraminifera (*Rotalina*), exactly of the same size and form as those liberated artificially from their parent, differing only in the second chamber beginning to show a yellow colour. "Here, therefore," he says, "we have a new proof that thé Foraminifera bear living young, and that these, at the time of their birth, are in a comparatively high state of development." [1]

Reproduction by gemmation is as probable in the Foraminifera as in *Difflugia*, in which latter it has been observed. When these animals are obtained in considerable numbers, the formation of colonies by gemmation (or budding) may easily be watched. This process takes place in the foot (*pseudopodium*), which gradually increases in size, acquires a nucleus and investing membrane, which ultimately separates and becomes a free animal.

The structure of the shells, in several different species, was the subject of careful investigation by Professor Williamson many years ago, and some interesting facts were then first brought to light. One of these series of facts serves to throw some light on the habits of the little-known animals which inhabit these shells. It was found on cutting sections, that the oldest cells have the thickest walls, and the youngest, or last-formed cells or chambers, the thinnest, and between these every gradation of thickness. This indicates successive depositions of calcareous matter on the exterior, for, if in the interior, the cavities would have been reduced and ultimately blocked up. The hypothesis suggested is thus stated:—"I have come to the con-

---

[1] "Ann. Nat. Hist.," vol. viii. (1861) p. 320.

clusion that the greater thickness of the earlier and older growths of the shell is owing to the circumstance that, on the addition of each new segment, the cell-wall by which it is enclosed has not terminated at the boundary of the individual segment itself, but has been prolonged over a considerable number, if not the whole, of the segments belonging to the outermost convolution of the animal. But before the newly-formed segment enclosed itself within the limits of its own calcareous shell, it must have had the power of spreading itself out in the form of a thin gelatinous layer, investing the whole of the preexisting organism."[1] That is to say, whenever the animal has so increased in size, in the last-formed chamber, that a new chamber must be constructed, the sarcode diffuses itself over the whole external surface of the shell, and deposits a new layer of calcareous matter over all the shell, then retreats within the last segment again, and completes a new segment, which becomes the last, and is the thinnest, because it has had no external stratum deposited upon it. Moreover, it is known that, at certain times, the sarcode *does* envelope the entire shell, and this explains wherefore it is done.

This hypothesis explains, and accounts for, the

---

[1] "Transactions of the Microscopical Society of London," vol. iii. (1850) p. 105.

increasing thickness of the walls of the various chambers, from the latest back to the earliest, the increased thickness being due to the successive deposition of thin external strata of calcareous matter, which strata would consequently be the most numerous in the oldest chambers, diminishing regularly and consecutively towards the most recently-constructed chamber. " Having so spread itself out as an investing body, it appears to have deposited new calcareous layers, covering over the greater part of the pre-existing external segments ; but in each layer so added open points have been left opposite to the mouths of the pseudopodian tubes in the layers previously formed, reminding us of the way in which similar apertures are left opposite the mouths of the canaliculi in membraniform bone-growths. In some instances, however, especially near the umbilical region, these tubes have been blocked up by the more recent investments."

We might also refer to the shell structure itself, and especially in those species in which "the outer layers are perforated in every direction by a network of minute anastomosing canals, which communicate freely with the exterior of the shell through numerous minute apertures." All these minute details of elaboration are marvellous, especially in the construction of the skeleton of animals so small, and of such simple organisation, that they are placed very near the bottom in the graduated scale of animal life,

> "How each fulfill'd
> The utmost purpose of its span of being,
> And did its duty in its narrow circle,
> As surely as the sun, in his career,
> Accomplishes the glorious end of his."

It would be almost hopeless to attempt any verbal description of all the various forms of shells which are constructed by the different species of Foraminifera.. Within certain limits, and in this instance very broad ones, each species has its own type of shell. This was more true in the days when the form of the shell was the principal character than it is at present, but there are some broad general features which find acceptance now. Primarily, there are three distinct varieties of texture—the *porcellanous*, resembling porcelain; the *hyaline* or *vitreous*, which are more glassy; and the *arenaceous*, or sandy. In the first the texture bears a strong resemblance to porcelain, especially when the surface is highly polished. Some of these are ribbed, others are channelled, whilst others are pitted either with small depressions or large areolæ. The vitreous shells have almost a glassy transparency, sometimes opaline, and these, again, may be sculptured on the surface with ridges or tubercles. Dead shells, when subject to long action of sea-water, lose their lustre and become opaque. Finally, the arenaceous are shells composed, partly at least, of particles of sand, obtained externally and agglutinated together by a cement supplied by the animal. The fine particles thus collected vary considerably, both in colour and substance. Sometimes the little particles are very uniform and methodically disposed; at others, they are less uniform, and form only an outer layer imbedded in a calcareous base. It seems a mystery how such a simple animal as that which inhabits these shells, little more than an atom of living jelly, can deposit one chamber upon another, in the multilocular shells, with the characteristic markings sculptured on the surface, or select and collect the minute particles of sand, and bind them together with admirable uniformity, like a miniature tessulated pavement, on the exterior of the arenaceous shells. Truly "there are more things in heaven and earth,"— and the deep sea also,—"than are dreamt of in your philosophy."

The simplest form of Foraminiferous shell is that in which the house is restricted to a single chamber. This may be as nearly as possible of a globose form, or oval, or attenuated at one end, so as to be almost pear-shaped, or the neck may be elongated so that it resembles a Florence oil-flask, some being plain, others ribbed or chequered. More complex forms consist of a series of spheres or ovals, adhering in a row, of which each successive chamber diminishes towards one end. In other forms the chambers approximate so closely that they are flattened at the poles, and united without any intervening isthmus. These may be straight or curved, but still gradually attenuated to one end. From the straight or curved forms the transition is easy to those in which the smallest end is curled into the commencement of a spiral. Then we encounter more perfect spirals, as complete as in the shell of a snail, but always with more or less distinct transverse lines, or furrows, indicating the division into chambers. Sometimes these spiral shells are nearly smooth, or they may be wrinkled, ribbed, indented, perforated, sculptured in

FIG. 5.—AGATHISTEGA.

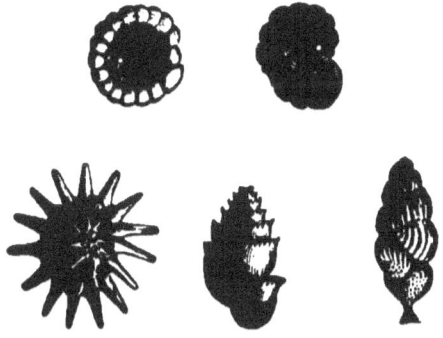

FIG. 6.—HELIXOSTEGA.

various ways, spurred, or keeled. In some the terminal opening is large, in others small. Then to all these may be added various irregular forms, long flexuous lobes packed side by side, or chrysalis-like forms with two or three longitudinal series of chambers packed closely like wedges, some regularly and others irregularly. In fact, there is such a multiplicity of forms (figs. 5, 6, 7, 8, 9) that they might be called polymorphous, and even a single species, as now understood, may include individuals which diverge remarkably from each other. It would be almost an exaggeration to say that no two individuals are exactly alike, but one is almost tempted to think so, in presence of such a multiplicity of form.

FIG. 7.—ENTOMOSTEGA.

One important result of deep-sea dredging has been to exhibit the relationship of the floor of the sea to the chalk deposit of the "dear white cliffs of Dover." "The dredging at 2,435 fathoms at the mouth of the Bay of Biscay," writes Sir Wyville Thomson, "gave a very fair idea of the condition of the bottom of the sea over an enormous area, as we know from many observations which have now been made with the various sounding instruments contrived to bring up a sample of the bottom. On that occasion the dredge brought up about 1½ cwt. of calcareous mud. There could be little doubt, from the appearance of the contents of the dredge, that the heavy dredge frame had gone down with a plunge, and partly buried itself in the soft, yielding bottom. The throat of the dredge thus became partly choked, and the free entrance of the organisms on the sea floor had been thus prevented. The matter contained in the dredge consisted mainly of a compact "mortar," of a bluish colour, passing into a thin—evidently superficial—layer, much softer and more creamy in consistence, and of a yellowish colour. Under the microscope the surface layer was found to consist chiefly of entire shells of Foraminifera (*Globigerina bulloides*), large and small, and fragments of such shells mixed with a quantity of amorphous calcareous matter in fine particles, a little fine sand, and many spicules, portions of spicules, and shells of Radiolaria, a few spicules of sponges, and a few frustules of diatoms. Below the surface-layer the sediment becomes gradually more compact, and a

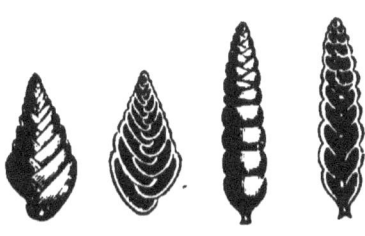

FIG. 8.—ENALLOSTEGA.

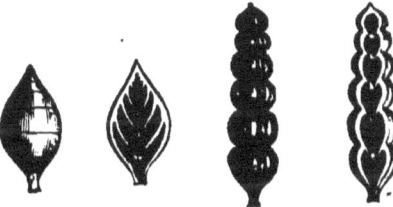

FIG. 9.—STICHOSTEGA.

slight grey colour, due, probably, to the decomposing organic matter, becomes more pronounced, while perfect shells of *Globigerina* almost entirely disappear, fragments become smaller, and calcareous mud, structureless and in a fine state of division, is in greatly preponderating proportion. One can have no doubt, on examining this sediment, that it is formed in the main by the accumulation and disintegration of the shells of *Globigerina;* the shells fresh, whole, and living, in the surface layer of the deposit, and in the lower layers dead, and gradually crumbling down by the decomposition of their organic cement, and by the pressure of the layers above; an animal formation, in fact, being formed very much in the same way as in the accumulation of vegetable matter in a peat-bog; by life and growth above, and death, retarded decomposition, and compression beneath." [1]

The foregoing extract, for which reason we have given it in full, shows some points of remarkable similarity between the Atlantic ooze and deposited chalk. There may be differences in chemical composition, but, when submitted to microscopical examination the resemblances are so great, or, as Sir Wyville Thomson says, "sufficiently striking to place it beyond a doubt that the chalk of the creta-

---

[1] "Depths of the Sea," p. 409.

ceous period and the chalk mud of the Atlantic are substantially the same." "Altogether, two slides, one of washed-down white chalk, and the other of Atlantic ooze, resemble one another so nearly that

FIG. 10.—GRAVESEND CHALK.

it is not always easy for even an accomplished microscopist to distinguish them."

In order to remove all doubt as to the impression he intended to convey, the same writer emphasises this point subsequently, when he writes:—" There can be no doubt, whatever, that we have, forming at the bottom of the present ocean, a vast sheet of rock which very closely resembles chalk, and there can be as little doubt that the old chalk, the cretaceous formation which, in some parts of England, has been subjected to enormous denudation, and which is overlaid by the beds of the tertiary series, was produced in the same manner, and under closely similar circumstances; and not the chalk only, but, most probably, all the great limestone formations. In almost all of these the remains of Foraminifera are abundant, some of them, apparently, specifically identical with living forms."[1] It was under these impressions, and with these views, that he publicly used the expression to which geologists took exception, that "we might be regarded, in a certain sense, as still living in the cretaceous period." But it seems that the mode of expression was censured rather than the opinion, since he afterwards declared that "the doctrine of the continuity of the chalk, in the sense in which we understood it, is now very generally accepted." In confirmation of this view may be cited the remarks of Professor Huxley, in his anniversary address as President of the Geological Society, in 1870, when he said, "Many years ago I ventured to speak of the Atlantic mud as 'modern chalk,' and I

---

[1] "Depths of the Ocean," p. 470.

know of no fact inconsistent with the view, which Professor Wyville Thomson has advocated, that the modern chalk is not only the lineal descendant, so to speak, of the ancient chalk, but that it remains in possession of the ancestral estate; and that from the cretaceous period (if not earlier) to the present day the deep sea has covered a large part of what is now the area of the Atlantic."

Accepting, therefore, this doctrine of the continuity of the chalk, and, consequently, its intimate relationship with our subject, let us endeavour to make some slight acquaintance with the microscopical characters of the common Kentish chalk. In order to do so, some little preparation is necessary. Take a small quantity of chalk in powder, say, two ounces, and place it in a glass bottle holding about a quart. Water is poured in upon the chalk until the bottle is nearly full. The whole is then shaken up, when it resembles milk in colour and consistency, and placed in some position where it remains undisturbed for half an hour. The heavier particles will have sunk to the bottom, the lighter remain suspended in the water, which latter is drawn off, and fresh clear water added, then the bottle may be ·stood aside to settle as before. This process is repeated again and again until the water is no longer turbid, after standing for a few minutes. The sediment, or deposit, at the bottom of the bottle will be found to be very much

less in bulk than the original quantity of chalk
experimented upon, but it will contain just what is
required for microscopical examination. If we take
a little of this sediment, and place it on a slip of
glass, then submit it to the microscope, we shall
find it to be composed almost entirely of delicate
little shells, those called *Globigerina* predominating.
These shells are carbonate of lime, easily dissolved
by acids, the empty, untenanted houses of Foraminifera. That which has been washed away in the
washing process consisted partly of the broken fragments of similar shells, and partly of amorphous
granules. Hence, therefore, the microscope teaches
that chalk consists, for the most part, of very minute
shells and fragments of shells, which were inhabited
by animals that lived and floated in the ocean thousands of years ago. By their identity of size and
form, it is not difficult to recognise in them the shells
of Foraminifera, belonging to precisely the same
species, in some cases, as those which are found
living at the bottom of the sea in the present day.

D'Orbigny computed that there were near four
millions of such little shells in one ounce of sand from
the Antilles. According to this computation a cube
of chalk of six inches in diameter in each direction,
and weighing sixteen pounds, would contain the
entire shells of not less than 1,024 millions of little
animals, and the broken fragments of nearly as

CHALK MAKERS, OR FORAMINIFERA. 55

many more. Such immense numbers are outside the range of our experience, and we can form no conception of them. Suffice it to say that one ounce

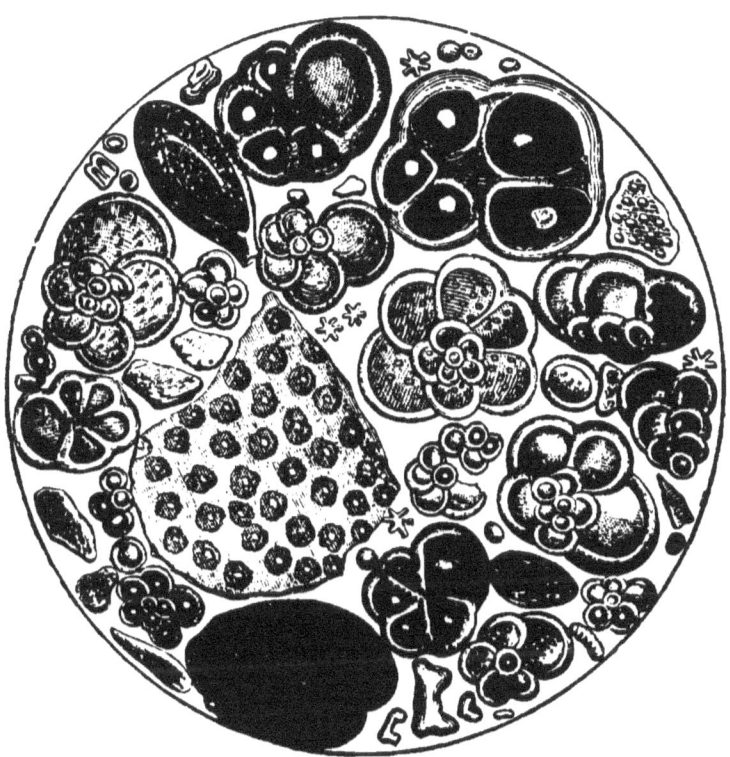

FIG. 11.—CHALK OF SICILY.

would contain the shells of more animals than there are human beings in the great metropolis of London. But what an infinitesimal fraction this ounce would be of the number of shells of these animals, living or dead, found scattered over the globe. Deep-sea soundings bring them up from the greatest depths that the line has yet reached, whole catacombs of them are entombed in the chalk and limestone rocks; they are scattered amongst the sand of the sea shore. In Europe and Asia together they cover thousands of miles. They may be found almost everywhere. The traveller who thinks he is out of their reach on the plains of Egypt is mistaken, for if he attempts to climb the pyramids of Ghizeh, he must do so by trampling on myriads of Foraminifera, imbedded in the nummulitic limestone at its base, which, could they but speak, might tell him of days,—long before the toiling Israelites were making bricks without straw,—of days long before Solomon was King of Jerusalem, or Elijah girded himself to run before Ahab to Jezreel (fig. 11).

In order to test the accuracy of other observers, as well as to procure some new determinations of the number of organisms approximately to be found in chalk, we undertook the experiments some years ago, of which the following is a summary:—An ounce of chalk, as taken from the pit, was subjected to the washing process already detailed, the lighter fragments being washed away, until a sediment was left of nearly pure foraminiferous shells. Half of this was cleaned, as much as possible, by boiling in caustic potash, and ultimately it was demonstrated that this supplied sufficient material to mount one hundred and ninety microscopical slides, of equal character, indeed the whole of that number were prepared, and compared. Each of these slides was estimated to contain one thousand shells, based upon the actual counting of two or three. So that, by calculation, it could be shown that in one ounce of chalk there were 400,000 shells.

Afterwards, and for greater security, another ounce of chalk was washed, even more carefully, and the calculation then showed upwards of half a million of entire shells, without reckoning the fragments which had been washed away, or probably the thousands that had been decanted off with the water in forty or fifty washings. Hence, it is evident that the maximum is not reached when it is affirmed that, at least, half a million of the shells of Foraminifera were contained in each ounce of chalk from that pit. The lump of chalk procured as the basis of these experiments weighed sixteen pounds, or 256 ounces, and consequently contained the shells of 128 millions of Foraminifera. Such a number is easy to name, but not so easy to imagine, a number which would occupy a person ten years to count, even if he could continue to count sixty per minute, for twelve hours daily.

Professor Ehrenberg, the celebrated German microscopist, calculated that there are one million and one-third of organisms in a cubic inch of chalk, and this calculation is, without doubt, very nearly correct. The block of chalk already referred to contained about 216 cubic inches, and, according to Ehrenberg's calculation, should contain 288 millions of shells. The total, according to our own estimate, based upon experiment, was 256 millions, these latter having been made before we were acquainted with Ehrenberg's figures. Two independent calculations, so nearly alike, must be held to strengthen each other, and seem sufficient to establish the fact that between one million and a quarter and one million and a third of foraminiferous shells are contained in each cubic inch of Kentish chalk

To convey some slight idea of the vast number of these little shells in an ounce of chalk, let us suppose for a moment that each shell was as large as the shell of the common garden snail (for it would appear to be near that size when seen through a moderately high power under the microscope). If such were the case, and these shells were placed side by side, then the half million of shells in one ounce of chalk would form an unbroken line of twelve miles in length. Or, if we take the whole of the shells contained in the experimental block of 216 inches (or six inches in diameter in each direction), and reckon them after the same rate, at 128 millions, then, on the supposition that each was of the size of a garden snail, and placed in a line, that line of shells would be 3,072 miles in length, and would occupy an express train seventy-seven hours to go from one

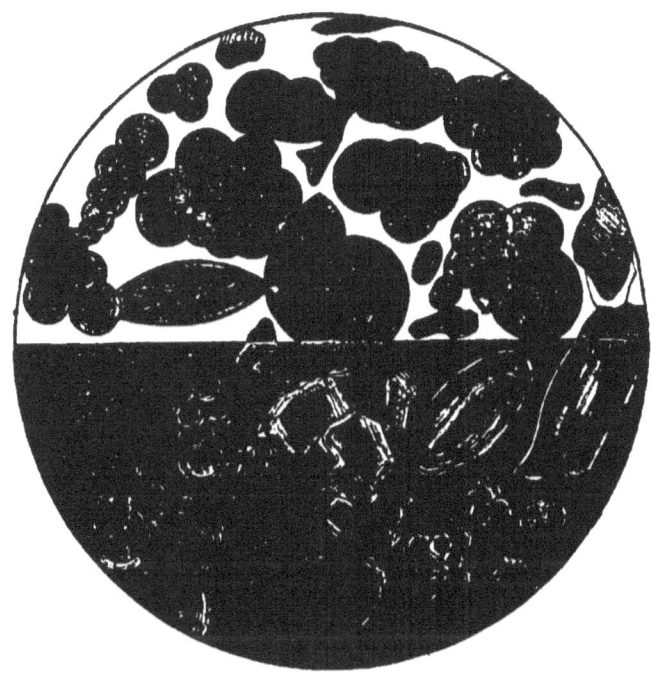

FIG. 12.—CHALK OF DENMARK.

end to the other, at the continuous rate of forty miles an hour.

It would be folly to attempt any calculation of the myriads of these little shells which are buried in the chalk of England alone, without any reference to similar beds in continental Europe (fig. 12). The

Kentish chalk pit whence our "lump" was derived, has furnished to the firm who now hold and work it, a total of more than half a million of tons, or 561,895 cubic yards of chalk. Figures would fail to convey any idea of the number of Foraminifera which this one firm of cement-makers have disturbed, and removed from their last resting-place. These minute shells are so small that it would require one hundred and fifty of many of them, placed side by side, to extend over one-twelfth of an inch. But if we take one of the most common forms (that of *Globigerina*), the diameter of which is the one hundred and fiftieth of an inch, it would require nearly ten millions placed end to end to reach a mile. Accepting this as the basis of another calculation, and we find that our Kentish firm have dug from their chalk pit the shells of Foraminifera—notwithstanding their minute size—sufficient to extend, at least, 1,006,915,840 miles. From one chalk pit, out of many hundreds in operation, one manufacturer of lime and cement has, within about a quarter of a century, dug out the shells of these minute animals, each one nearly invisible to the naked eye, in sufficient quantity to reach more than a thousand millions of miles. Or, to represent it under another form, enough to go round the world, in an unbroken line, more than forty thousand times.

We have ventured the assumption, that, taking

into account the past as well as the present, the Foraminifera are the most ubiquitous and numerous of all created things. That, living or dead, themselves, or their shells, far exceed in number all the visible evidences we possess of any other kind of organism. A few suggestions is all we would venture, but from these it will appear that our assumption was not a reckless ône, and not altogether without the basis of probability.

Already we have hinted at the universal diffusion of living Foraminifera over the sea bottom, down to the lowest depths, from the poles to the tropics, or at least as near to the pole as the absence of ice would permit the condition of the sea bottom to be ascertained. When it is remembered that the ocean covers nearly three-fourths of the surface of the globe, it must be admitted that the available space for the distribution of living Foraminifera is extremely large. The result of all the dredgings and soundings hitherto made has been the ubiquitous presence of these minute animals. Drawings made of the magnified portions of the sea bottom, attached to the "Narrative of the *Challenger* Expedition," serve to corroborate this fact, against which there is no substantial evidence in any direction, so that it may be accepted as a "fact incontrovertible," that the sea bottom, at all depths, independent of latitude and longitude, is 'inhabited by living Foraminifera. From this it may, at least, be inferred that one-half of the surface of the globe is still inhabited by the members of this cosmopolitan family. When it can be written of one genus that "it is an inhabitant of all seas, through a great range of depths," and again, that "it has been found in abundance in every maritime region,"[1] we may fairly conclude in favour of the presence of Foraminifera in all seas. And when we find it recorded that another genus constituted 97 per cent. of soundings brought up from a depth of 2,000 fathoms in the mid-Atlantic, and a like proportion at depths of 1,260 and 1,607 fathoms, not far from Greenland, it may be inferred that some forms of the Foraminifera are exceedingly abundant, to use no stronger terms, on the deep-sea bottom. Hence we consider ourselves justified, without multiplying details, in declaring that living Foraminifera are plentiful, all over the sea bottom, at all depths, and in all latitudes.

In the next place, we may endeavour to obtain some idea of the immense number of shells and casts of Foraminifera that have existed in the past. The largest forms are those of the Nummulites, which, with some intermixture of other types, constitute a stratum of limestone, "not unfrequently attaining a thickness of 1,500 feet, which extends in an east

---

[1] Dr. Carpenter's "Introduction to the Foraminifera," p. 178

and west direction, through Southern Europe, Libya and Egypt, and Asia Minor, and is continued through the Himalayan range of Southern Asia, into various

FIG. 13.—CHALK OF MEUDON.

parts of the great Indian Peninsula, where it acquires a very extensive development."

The "Nummulite Limestone" of Alabama extends over an immense area in that state, and is almost entirely made up of the remains of one species of *Orbitoides*, which is found also in Madagascar and the West Indies.

Professor Baily remarks that Charleston is built upon a bed of several hundred feet in thickness, every cubic inch of which is filled with myriads of perfectly preserved shells of Foraminifera.

The most prolific beds of defunct Foraminifera are those of the chalk, which have a considerable extent in the south of England. Accurate measurements of the thickness of the chalk have not been made, but Sir Henry de la Beche estimates the average thickness at 700 feet. The average thickness of the chalk on the Sussex coast is estimated at 800 feet. The flinty chalk at Dover is 350 feet in thickness. At Diss, in Norfolk, the thickness was ascertained by boring to be 510 feet. The number of square miles of chalk in England must be very considerable, not to mention its continuation in Northern France (fig. 13), and other deposits in different parts of the continent. No human calculation could embrace the myriads of shells of Foraminifera which these deposits contain, and yet they are only a few of the most important accumulations of these remains. We must leave imagination to fill in the details, on the basis of these suggestions.

Apart from their natural position, whether living at the bottom of sea, or dead and buried in the various deposits, of which they form so important a part, it is remarkable how we are always accompanied through life by these fossilised remains, distributed by human agency. If we realise the fact that the chalk is composed of them almost entirely, then we must admit that wherever we encounter chalk we have Foraminifera. Mixed with sand and converted into mortar, the calcined chalk in the form of lime must still contain myriads of them, so that not only every brick building, but the plastered walls and ceilings are mausoleums of hosts of Foraminifera; whitewash abounds with them; we swallow them in chalk mixtures whenever our own internal arrangements are disturbed; glazed cards are manufactured by their agency, and transferred to our pockets; crayon drawings are indebted for their high lights and broad effects to shell remains; and, indeed, it is more difficult to determine where no trace of them can be found, than to enumerate their artificial localisation by the hand of man. Through untold ages myriads of minute atoms of life have built their tiny shells, and died—and wherefore

"They roam'd, they fed, they slept, they died, and left,
Race after race, to roam, feed, sleep, then die,
And leave their like through endless generations."

CHAPTER III.

LATTICE WORKERS, OR POLYCYSTINA.

IT has come to be a firm belief with some people, and not without reason, that no natural object is wholly devoid of beauty; that, to put it in other words, there is no natural object which is not to a more or less number of intelligent, reasoning, and reasonable beings possessed of beauty. There may be cases, as for instance, with reptiles and toadstools, that the number is very limited of those who recognise beauty in what the great majority consider but too horrible and disgusting. Take it for granted that the admiring few may be very few indeed, still they will be, after all, that few to which the objects in question are best known, and known clear from all prejudice or bias. Whether there is any analysis of beauty which makes it to consist in fitness, or whether it is capable of analysis at all, does not now belong to us to inquire, since we are not about to plead specially on behalf of any objects generally excluded from the circle of the beautiful.

There are yet other objects which, either in virtue of their form, colour, or combination of both, at once make their impress on all observers, and at once force the feeling, if not the expression in words, O how beautiful! Both classes of objects occupy their allotted places in creation, and both seem equally fitted for that position ; yet, when viewed apart from their associations, when appraised according to their merits as isolated objects, how vast a difference in the result of the poll. In one case the verdict is almost unanimously, Aye, in the other it is almost as unanimously, Nay. Amongst all the myriads of minute objects for the microscope, contained in the cabinets of the curious, there are none more generally attractive, more popular, or when effectively illuminated, more beautiful than those little skeletons or deserted houses, which are known by the name of *Polycystins.*

Even now this strange name may not be familiar to some who are scanning these pages, and the objects which that name is supposed to represent may be as strange and unknown too. Under such an assumption it may be well to explain, that many years ago a sort of chalky earth was found in various localities [in Barbados, and some of it was sent to Europe as containing many beautiful and interesting objects for the microscope. The most numerous of these objects contained in the Barbados earth were the Polycystins, as they came to be called, which were so minute as to be invisible to the naked eye, except as small grains of dust, but which under a low power of the microscope, are seen to be a variety of elegant forms, or skeletons, perforated with holes so as to resemble lattice-work, composed of a white translucent substance resembling flint, and none of them so large as a pin's head.

It was apparently in January, 1846, that Dr. Davy, who was then resident at Barbados, first detected these organisms in the chalk beneath the coral rock, and when Sir R. Schomburgh went thither he was informed of the fact, and some of the skeletons were exhibited to him under the microscope. Full of enthusiasm and delight at the discovery, a portion of this deposit of chalk marl was sent to the eminent Professor Ehrenberg, of Berlin, and during the same year this microscopist announced the discovery, in a communication to the Berlin Academy of Sciences, in which he said, " for these organisms constitute part of a chain, which, though in the individual link it be microscopic, yet in the mass it is a mighty one, connecting the life-phenomena of distant ages of the earth, and proving that the dawn of organic nature co-existent with us, reaches farther back in the history of the earth than had hitherto been suspected. The microscopic organisms are very inferior, in individual energy, to lions and elephants, but in their united influences they are far more important than all these animals."

Yet this was not the first time that these minute animals, or rather their skeletons, became known, for already previous to 1838, a few fossil forms had been observed, and in that year these were grouped together by Ehrenberg under the name of *Polycystina*, which they still partly retain. Prior to the discovery of the Barbados earth these organisms had been found in the chalks and marls of Sicily, Oran, and in Greece, and sparingly in the Tripoli of Richmond in Virginia, and in Bermuda.[1] Later on they have been brought up in profusion, in a living condition, from out of the depths of the sea. Our whole knowledge of them may be circumscribed within the limits of the last half century.

It would be an interminable task to enumerate and describe all the varied forms of these skeletons or shells, (Plate II.), but Mrs. Bury has given a characteristic introduction to their external appearance :—" Among the most striking differences of form," she says, " may be mentioned that some are like numerous little globes growing out of each other, beginning with a small one at the apex, and each increasing in size as

---

[1] A *résumé* of what was known of them, in their relation to existing animals, is contained in " Microscopical Siliceous Polycystina of Barbados," by Sir R. Schomburgh, in " Annals of Natural History," vol. xx. p. 115. 1847.

they descend, sometimes so gradually as to look like storied beehives piled up one on the other. In others, the globes or bells dilate greatly, and rapidly, towards the base, till the lowermost exceed in width the most outrageous crinoline! Then there are single round balls pierced through and through, with spikes sticking out in all directions, but ever pointing from the centre. Again, to compare small things with great, there are shapes of the Egyptian pyramids, and others stretched out and narrowed into obelisks. Different from all these are very numerous flattened disks, which appear to grow in concentric circles, some becoming bordered, others spiked round the edges, and many having very extraordinary radiating arms in endless variety."

" A very remarkable feature in the Polycystins are their exuberant outgrowths. Sometimes they are merely spines, projecting in a tolerably regular and always radiating manner ; but sometimes these spines or projections branch out and subdivide in the most whimsical arborescent forms, so as to assume the shapes of stag's antlers, or even the more complicated delicate branchings of the once famous bedeguar of the rose. Apparently these spines, whether simple or compound, always are to a certain extent hollow,[1]

---

[1] " The appearance of tubularity is an optical illusion, engendered by the longitudinal ribs, of which the elongated spines

so as to have been permeated by the sarcode substance during growth; sometimes, instead of becoming finely attenuated, they become bulbous at their points, and these bulbils swell, become cellular or foraminated, and assume very much the appearance as if they were buds from the parent, and intended to break forth and commence life as fresh individuals."
"Very remarkable instances are seen in those forms in which a central spheroid body is enveloped by a sort of fine siliceous web or sponge, which gradually breaks away, as the centre sends forth stalks (three to six, but normally four) celled and chambered as complicately as any foraminifer; and pushing their way through the sponge-like envelope, and beyond it, their ends become club-shaped, often strongly spiked at the extremity, but the swollen part, containing what looks like a reduplication of the central parent form; and these seem as if they might possibly break away, and become in their turn centres of growth."

"But though some are symmetrical, or nearly, they are by no means universally so : in fact, the greater number display the most grotesque polymorphism. For instance, you find one with the flowing outline of some elegant Etruscan vase, with a tapering base; the next specimen you see may have ugly nose-

---

may be said to be made up."—Wallich : "Ann. Nat. Hist" (1864), p. 81.

shaped handles, adhering to its sides in a clumsy way; in another example the nose-like protuberances enlarge, thicken, become wrinkled, and, however droll-looking, militate sadly against one's ideas of elegance. Yet look again and again, and in some favourably well-grown individual, lo! there are the clumsy excrescences lengthened out, curved, refined and developed into supports, worthy of the famed Delphic tripod of the Pythian priestess; the gently swelling ends of the feet, meanwhile, indicating appearances as of little reservoirs, or deposits of material, accumulating for future use, either as buds, or for some further development. Again, in some of the pyramid or obelisk shapes, the fenestræ, small at the apex, widen more and more towards the base, and a fine inner lattice work is seen to line these too-wide-open windows. But the variations in form assumed by these cunning artificers are far too numerous to mention; and although they do not work by mathematical rules and compasses, as has been sometimes represented, they have within themselves a mysterious unerring rule, which guides every thread, every particle of internal sarcode, or external silex, into the position, shape, and size, best suited to the situation, surrounding circumstances, and requirements of each individual organism."[1]

---

[1] "A Popular Description of Polycystins," by P. S. Bury, in *Science-Gossip*, May 1, 1865.

Assuming, then, that the skeletons of the animals called *Polycystins*, which are found in various deposits as fossils, as well as in the mud of the ocean at the present day, are minute siliceous, or flinty, objects of very variable shape, perforated for the most part with numerous openings, as variable as the contour of the skeleton, and that these skeletons are sometimes symmetrical and sometimes not, it will now be our object to clothe these skeletons, and ascertain what is known of the living animal which constructed these glass houses, or flinty tenements, in which they dwelt. In this instance, as in some others in the present volume, we accept the term *sarcode* as the representative of the simple, glairy, gelatinous flesh of the animal, however modified, but in order to avoid confounding this substance with *flesh*, as we understand and recognise it in the higher animals, it is better to employ a different term, and we have it at hand, as already adopted and recognised, in *sarcode*.

There is much in common in this respect between the Foraminifers and the Polycystins, for Dr. Wallich, observes:—"I believe that between the degree of differentiation of the sarcode body observable in the Foraminifera, and the Polycystina, no difference of importance really exists. In both families the pseudo-

---

[1] "On the Structure and Affinities of the Polycystina," by G. C. Wallich, M.D., in *Quart. Journ. Microscopic Science* v. p. 67. 1865.

podia are given off principally with reference to the number and position of the apertures occurring in the shell. In both coalescence of the most complete kind takes place immediately on the escape of the sarcode stolons through the main, or secondary apertures, to such an extent as occasionally to constitute a delicate externally investing layer, between the inner surface of which, and the outer aspect of the shell, the process of mineral deposit goes on."

In order to avoid needless repetition we would refer to what we have written of the animal of the Foraminifera, which corresponds so nearly with that of the Polycystina. For general purposes it would be sufficient to say that the animal is, practically, somewhat of an *Amœba*, or *Actinophrys*, enclosed in a shell or skeleton ; and that an *Amœba* is one of the simplest forms of animal life, consisting of a single minute nodule of crystalline jelly without organs of any kind. Imagine a minute speck of isinglass jelly endued with life, and you have an *Amœba* or *Proteus*. It has no mouth, no eyes, no legs, no hands, no feet, no stomach ; yet it performs the functions of all these. No mouth, yet it extemporises a mouth from any portion of its surface, involves, and draws in its prey ; no legs, yet it elongates, and pushes out, any portion of its gelatinous substance like a leg to move itself about ; . no stomach, yet it takes its food and gathers it into the midst of its gelatinous mass, digests it, and rejects the undigested portions. Now it is almost spherical, then oblong, lengthens itself into a long and narrow body, and then as speedily becomes triangular, starshaped, or many-sided. Every instant changing its form and assuming every shape of which a plastic lump of gelatine is capable. In the simple *Amœba*, portions are thrust out and extemporised into legs or arms, and drawn back again and absorbed when their work is performed ; but in a slightly advanced animal, called *Actinophrys*, these *pseudopodia*, or false arms and legs, are more thread-like and more permanent, as they are also in the Polycystins.

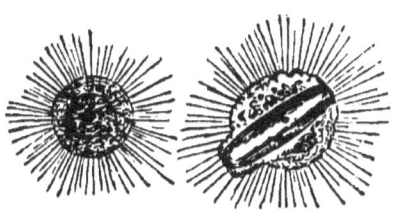

FIG. 14.—ACTINOPHRYS.

The bulk of the sarcode occupies the interior of the shell, or skeleton, whilst the pseudopodia are protruded through the many orifices, and seem to coalesce at the base, and flow over the skeleton, so as to form a thin layer like a pellicle enclosing it. By means of these protruding arms the animal not only moves from place to place, but is enabled to procure its food (fig. 14).

Adverting for a moment to the little animal already alluded to, which is almost as commonly found in fresh-water as the *Amœba*, and bears a close relationship to it, called the Sun-animalcule, or *Actinophrys*. It is through this little creature that some knowledge may be obtained of the animal of the Polycystins; indeed it might almost be termed a Polycystin without a shell. The body is spherical, and is of that peculiar jelly-like consistence already described in the *Amœba*, from which, however, it differs in not being so variable in shape, not suffering such continual changes, but retaining its globose form and also in the more slender threadlike pseudopodia which radiate from the body in all directions. Dr. Carpenter remarks that "in Actinophrys the pseudopodia are very numerous, and, when fully extended, are long, slender processes, that gradually taper from base to point, and issue from the body in a radial direction; they generally remain distinct, when they come into mutual contact, never undergoing that complete fusion which is common in those of the Foraminifera; and a slow movement of granules may be seen to take place along their margin, when the observation is continued for a sufficient length of time, under high magnifying powers." Although several memoirs[1] have been written, dealing more or less with these organisms, their history on many

---

[1] Kolliker: "The Sun Animalcule." *Quart. Journ. Micr. Sci.*, vol. i. (1853), pp. 25, 98. Claparède, on "Actinophrys," in Annals Nat. Hist.," vol. xv. (1855), pp. 211, 285.

points still remains very obscure. All that is needful for the present purpose may be contained in a brief summary : " The Actinophrys seems to have little or no power of moving spontaneously from place to place, and it obtains its food entirely through the instrumentality of its pseudopodia, which, by their peculiar adhesive property, attach themselves to bodies that come into contact with them. Not only motionless particles of vegetable matter, but actively-moving Infusoria are thus entrapped. When the prey is large and vigorous enough to struggle to escape from its entanglement, it may usually be observed that the neighbouring pseudopodia bend over, and apply themselves to the captive body, so as to assist in retaining it, and that it is slowly drawn by their joint retraction towards the body of the Actinophrys. In other cases, however, the captive seems as if it were paralysed by the contact of the pseudopodium, remaining motionless for some seconds, and then, without any visible movement of its captor, gliding either slowly, or rapidly in a centripetal direction, along the margin of the pseudopodium to which it adheres, until it becomes jammed, as it were, between the base of this and a neighbouring one. It is usually, in fact, by thus gliding along the margin of the pseudopodium, as if propelled by an invisible peristaltic contraction of its sarcode, rather than by a visible retraction of the pseudopodium, that any small body, not capable of offering active resistance, is conveyed to its base. Now and then it seems as if the appetite of the Actinophrys were sated, or the prey not approved of, for after a few seconds the movements of the latter feebly recommence, and it glides off the pseudopodium, without any effort on the part of the Actinophrys to retain it. When, on the other hand, the captive is to be used as food, it becomes invested by an expansion of the protoplasmic substance which the body of the Actinophrys sends forth on either side of that of the captive, so as to meet and enclose it ; and thus a marked prominence is formed, which gradually subsides as the food is drawn more completely into the interior. There can be no doubt whatever that aliment may be thus ingested at any part of the surface, a new mouth, so to speak, being extemporised whenever and wherever there is occasion for it. The struggles of the larger animals, and the ciliary action of the Infusoria, may sometimes be observed to continue even after they have been thus received into the body ; but these movements at last cease, and the process of digestion then begins. Whilst the digestive process,—which usually occupies some hours,— is going on, a sort of slow circulation takes place in the entire mass of the sarcode. If, as often happens, the body taken in as food possesses some hard

LATTICE WORKERS, OR POLYCYSTINA. 79

indigestible portion, this, after the digestion of the soft parts, is gradually pushed towards the surface, and at last escapes by an anal opening, which extemporises itself for the occasion.[1] The reproductive process in Actinophrys is still in need of further elucidation. Binary subdivision is one method, in which the body is constricted all round, which constriction gradually deepens, until the connecting band becomes thinner and thinner, and complete separation takes place. This process occupies about half an hour. It has also been affirmed that the junction of two individuals has been seen to take place, which has been assumed to correspond to conjugation in certain Algæ, but there is not sufficient evidence to associate this union with an act of reproduction. Other appearances have also been detailed, from which it has been inferred that true sexual products are formed in the interior of the bodies, and that spermatozoids are liberated and ovules fertilised. " Nicolet has stated that in Actinophrys the generative organs consist of a central spherical membrane, enclosing little globules, which are the rudiments of 'eggs,' surrounded by a 'gelatinous granular layer,' the granules of which appear to be the reproductive organs," but this view is doubted, and has not been substantiated.

---

[1] Dr. Carpenter : "Introduction to the Foraminifera," p. 19.

Although, in some minute points of its history, the naked Actinophrys is not identical with the shell-bearing Polycystins, the knowledge of the one may be helpful to the knowledge of the other, and it must be remembered that it is only in recent times, consequent on great improvements in the microscope, that any real history of the structure and habits of such minute organisms has been traced, and in this direction much still remains to be accomplished. Difference of opinion as to the interpretation of appearances still prevails, and an instance of this kind occurs in regard to the inception of food, for, on this point, Dr. Wallich says, that "during many years' study of the Foraminifera, Polycystina, and all pelagic Rhizopods, in their living condition, notwithstanding a keen look-out for an example, he has invariably failed to discover a single instance in which there was satisfactory evidence that solid matter had been taken into the substance of the body as food. This fact derives additional weight from the circumstance that some of these Rhizopods occur at times in immense numbers. It is difficult, therefore," he thinks, "to conceive how all should present themselves devoid of everything like solid incepted matter, were such matter essential to their existence. For it must be manifest that, as generally attributed to the Rhizopods, the processes referred to partake of the miraculous, and, what is particularly notable, that it is not amongst the highest members of the class that these processes seem to be carried on, in the absence of organs wherewith to effect them, but in those lower types in which we have hitherto failed to detect a trace of such organs."[1] Whether solid food is absorbed or not, of a size sufficient to be distinguished, is of little consequence to this our history, and having stated both sides of the case we will proceed, with the intimation that even those who doubt the absorption of solid food particles, by either the Foraminifera or Polycystina, admit that Amœba and Actinophrys " do unquestionably incept solid food particles, digesting and assimilating whatever portions are nutritious, and extruding the remainder."

There seems to be a general agreement that a kind of imperfect granular circulation takes place in all these allied forms, not only in the central sarcode, but along the extensions, in the form of pseudopodia. This " streaming of granules," whether in the body or in the pseudopodia, is not admitted to be referable to a vital power in the sarcode, in which the primary vital function consists in the contractility of the substance, but rather that the progression of the molecules is the exponent of that function.

From incidental remarks, in the course of his

---

[1] *Quart. Journ. Micros. Science*, 1865, p. 65.

"Introduction," it may be inferred that Dr. Carpenter regarded the "cyclosis," or flowing of granules, in the body, or extensions of the body, in the Foraminifera and Polycystina, as something more than mechanical, and of the nature of vital action. On the other hand, Dr. Wallich maintains that the movement is one of a purely mechanical nature, in terms which are emphatic and not liable to be mistaken. "I feel bound to express my conviction," he says, "that such cyclosis is in no way to be regarded as an independently acting vital function, resident either in the protoplasm proper, or in the granules suspended within it; but is a purely mechanical result affecting the granules, only through movements executed by the vital contractility, which is an inherent attribute of animal protoplasm. When these movement cease the circulation of granules ceases; when they are resumed the circulation of granules is resumed also. Notably in the Foraminifera and Polycystina, these movements become manifest only to the extent of causing a nearly constant efflux and influx, in opposite directions, of the protoplasmic matter entering into the formation of the pseudopodia, and also in that portion of it which, in most cases, constitutes a delicate investing layer on the exterior surface of the shells. It is, however, in the naked Amœbæ that the pseudo-cyclosis attains its most energetic and characteristic limit, and we can most readily perceive that it is produced by a mechanical and not a special vital agency."[1]

Up to the period of the publication of the *Challenger* reports, the following was considered as the sum of our knowledge of the reproduction of the Polycystina:—

It was not assumed that we had any evidence in favour of sexuality in the reproduction of these little animals, or, in fact, any supposition of sexuality in a group of such simple organisation as that to which they belong. There was, nevertheless, reproduction after a distinct type, associated with what have been termed the "yellow bodies" of the sarcode. Both in the animal of the Foraminifers and of the Polycystins these "yellow bodies" are found, to which Dr. Wallich was the first to apply the name of *sarcoblasts*, with the view of distinguishing them from other corpuscles of a similar appearance. These, he contends, are the true rudiments of the young Foraminifera and Polycystina, and probably of all Rhizopods. "And whereas," he says, "in the case of the marine and fresh-water genera, I have been enabled to collect sufficient data to prove that these bodies, although developed within the parent protoplasm, become ultimately extruded therefrom, and

---

[1] "On the Rhizopoda," by Dr. G. C. Wallich, in *Monthly Micro. Journ.*, vol. i. (1869), p. 233.

in the testaceous (or shell-bearing) forms, that the deposition of the shell material dates, as a general rule, from this period ; the development of the ' testa,' while still within the parent sarcode, occurs in some of the Foraminifera, and brings the phenomenon within the category of viviparous reproduction."[1] These *sarcoblasts* are very conspicuous in the Polycystins, and are more readily seen than in the Foraminifera, in consequence of the crystalline character of the skeleton. They occupy a position for the most part immediately within the silicious framework.

No better illustration of Dr. Wallich's views on this, which was then considered an important phase of the life history of the Polycystins, can be given than in his own words: "In 1863," he writes, " I pointed out, for the first time, that the so-called ' yellow bodies,' as also more or less perfectly colourless, but in other respects perfect homologues of them, are present throughout *all* the Rhizopodal families, both oceanic and fresh-water; this statement being based on long personal experience. At the time indicated, they had been observed by others only in three pelagic families, their brilliant yellow tint in the pelagic families being regarded as their distinguishing

---

[1] Dr. Wallich : *Quart. Journ. Micr. Science* (1865), p. 71. Afterwards Cienkowski in 1871, and Brandt in 1881, have shown that the "yellow cells" do not belong to the Radiolarian system, but are symbiotic unicellular Algæ.

character, until it was shown by me that absolutely identical bodies, in all save colour, are common to the entire class. Their office had, moreover, until then been either altogether unrecognised, or, so far as I am aware, referred to only more or less incidentally, as in some manner connected with reproduction. It was in the course of a laborious day-by-day series of observations on the fresh-water and littoral Rhizopods, that I was enabled to compare and trace clearly, and consecutively, the mode of origin of these remarkable bodies, and to prove beyond all reasonable doubt that they constitute a true reproductive organ, formed either *directly* by the aggregation into minute sphæroidal masses of granular, probably germinal, particles, which, up to the period of this change taking place, are more or less uniformly distributed through the sarcode mass generally; or, *indirectly*, by the subdivision of the contents of the granular nuclear mass itself, without, however, acquiring in any instance a membranous covering. For these combined reasons, which had obviously made the term 'yellow cells' a dangerous misnomer, I designated them *sarcoblasts*. Whether in the fresh-water littoral or oceanic Rhizopods, the sarcoblasts *invariably* constitute, when liberated from the parent organism, either at once the infant shell-less organism, or the *nidus*, and at the same time the infant mass of sarcode, within which the rudiment of the shell, or other mineral framework, of the organism is secreted. In the Foraminifera and the Polycystina they are the *nidus* within which (in the Foraminifera) the primordial calcareous chamber of the shell is secreted, or (in the Polycystina) the earliest siliceous rudiment of the siliceous framework, or perforated shell."

The "sarcoblast" is in reality a small sphæroidal mass of sarcode, generally varying in size, in the different families, within certain limits, and wholly devoid of any true membranous or other special covering. When first observable, it consists of an almost colourless, hyaline, viscid, and basal fluid, of the consistence and appearance of the white of an egg, within which are distributed peripherally, but without any approach to regularity, a number of granules of more consolidated, as well as more or less faintly coloured sarcodic substance, and invariably (in the case of the sarcode of the oceanic Rhizopods) of a tolerably brilliant yellow tint. But, by carefully focussing down to an equatorial plane, a central portion, altogether devoid of granules, and occupied solely by the pure basal sarcode, may quite readily be detected. It is to this clear portion that the somewhat misleading name of *nucleus* has been applied ; perhaps owing to the idea that its apparently higher refractive power, as compared with the surrounding mass, may be due to its being a specialised product, or, in other words, not identical with the sarcode of the rest of the sarcoblast. Through some subtle reproductive operation (of which we have learned nothing from actual observation, in consequence of the minuteness of the particles concerned) the yellow granules become, after a time, collected together into the sphæroidal masses, now become sarcoblasts; and, possibly, two sexual elements are, at this stage of the organism's history, brought into contact. It will, I hope, be clearly understood, that I throw out this view simply as a surmise, resting on no more stable basis than a fact, observed by myself (for the accuracy of which I am ready to vouch with perfect confidence), that the so-called nucleus of the sarcoblast becomes eventually, on the escape of that organ from the parent structure, the active centre of shell or skeleton development.[1]

To summarise what we have narrated, and quoted, of the reproductive process in these organisms: the sarcoblast, or embryo Polycystin, is first recognised, in the body of its parent, as a minute, spherical, or almost spherical, speck or lump of granular sarcode, yellowish in colour, and variable in its diameter, as soon as it becomes free, ranging, perhaps, from five to twenty micro-millimetres, or from one five-thousandth

---

[1] "On the Radiolaria as an Order of Protozoa," by Dr. G. C. Wallich, in *Popular Science Review*, new series, vol. ii. p. 273.

to one thirteen-hundredth part of an inch, that is, so minute that one of medium size would have a diameter of about one three-thousandth part of an inch, so that three thousand of them arranged in a line would only extend an inch ; that these minute embryos are destitute of anything like a proper cell-wall, but soon after expulsion from the parent exhibit a clear spot, which might casually be mistaken for a nucleus, but which in reality is the rudiment of the future skeleton ; these specks, or "yellow spots," being seen resting in a sort of irregular layer immediately beneath the flinty framework, or skeleton, of the parent, and subsequently cast off through one of the numerous openings of the shell. There is no doubt about the rudimentary condition of knowledge of the development of the young Polycystins from the sarcoblast, for even as much as this is admitted by Dr. Wallich. "Occasionally," he says, "during calms within the tropics, the sarcoblasts of the Polycystins, and other oceanic Rhizopods, may be taken in immense numbers, although, owing to their extreme minuteness they are easily overlooked. The profusion, however, in which they occur, in every stage of growth, affords the means of tracing their history in al its consecutive phases, and it is highly desirable that they should be carefully collected, and studied by all who enjoy opportunities of obtaining them, in their normal condition."

The method by which lime is deposited by the animals of the Foraminifera, so as to construct their calcareous shells, and that by which flint is collected, in the allied group of Polycystina, for the elaboration of their siliceous skeletons, appear to be the same in both cases of all the essential points, although the material is different. The structures are formed in both by deposits at right angles to the principal line of growth, and in one direction only.[1] Of course, the material for this operation has to be derived from, and is held in solution by, the water of the ocean, whether it is to be regarded as a true secretion, or merely an exudation, is a subtlety which need not trouble us ; it is sufficient to remember that the development and extrusion of the sarcoblast precedes the first appearance of the embryonic skeleton. " All subsequent deposits of silex, or flint, whether in the shape of foraminated chambers, or spines, or other portions of the structure, take place on the same plan—namely, by the deposit of silex at right angles to the axis of growth, of the part immediately in question. In the formation of the chambers the deposit usually goes on from a number of points simultaneously around the free margin, the points

---

[1] "Remarks on the Process of Mineral Deposit in the Rhizopods," by ,Dr. G. C. Wallich, in "Annals of Natural History" (Jan. 1864), p. 72.

becoming filaments, and the adjacent filaments ultimately anastomosing, or rather coalescing, as soon as they come into contact. The spines are never tubular, the appearance of tubularity in the spines of some genera being due to the existence of short longitudinal furrows and buttresses on their inner aspect, where generally may be seen an aperture around a portion of the margin of which the base of the spine has taken its rise. When loops or festoons occur the process is still the same, as these may be seen in every stage of growth, from the first projection of a minute filament to the stage at which the coalescence would have become complete had the protecting and formative living sarcode been left to fulfil its office. In short, the process may be familiarly likened to that by which the glass-worker extends his plastic and half molten material from point to point, when manufacturing a miniature basket work. Of course the thickening of each portion is by subsequent deposit around the original thread."[1]

The number of genera and species of Radiolaria was increased enormously by the results of the *Challenger* expedition. The total number described by Haeckel, in his " Report," was 739 genera, and 4,318 species, of these 3,508 were new, as against 810 pre-

---

[1] Dr. Wallich, in *Quart. Journ. Micro. Science*, v. p. 82.

viously described, and yet the riches of the *Challenger* collection was by no means exhausted.

From this time the old name of Polycystina came to be absorbed in that of Radiolaria, and the old opinions of structure were modified into something like the following summary of the views propounded by Professor Haeckel.

The Radiolaria are marine rhizopods, whose one-celled body always consists of two main portions, separated by a membrane,—viz., an inner central capsule, with one or more nuclei, and an outer capsule (extra-capsulum), which has no nucleus, and the pseudopodia, the central capsule, is, in part, the general central organ, and, in part, the special organ of reproduction. The outer capsule is partly the general organ of communication with the outer world, and partly the special organ of protection and nutrition. The majority also develop a skeleton, which presents the utmost variety of form, generally composed of silica (or flint), but sometimes of an organic substance. The Radiolarian cell usually leads an isolated existence, but in a small number of instances the one-celled organisms are united in colonies.

It is contended that the special peculiarity of the one-celled organism, by means of which it is distinguished from all other rhizopods, consists in its differentiation into two separate constituents (the central capsule and the outer capsule) and the formation of a special membrane which separates them. The outer capsule is usually more voluminous than the central capsule or inner portion. The protoplasm of the former is emphatically different from that of the latter. The central capsule is, on the one hand, the general central organ for the discharge of sensory and motor functions, and, on the other hand, the special organ of reproduction. The outer capsule is not less significant, since, on the one hand, it acts as a protecting envelope to the central capsule, as a support to the pseudopodia, and as a foundation for the skeleton or shell, and, on the other hand, its pseudopodia are of the utmost importance, as peripheral organs of movement and sensation as well as of nutrition and respiration. Hence the central capsule and the outer capsule are to be regarded as the two characteristic related parts of the one-celled organism.

The central capsule is originally a spherical body, which is separated from the peripheral portion by an independent membrane, which latter appears early, and is in most cases persistent through life. The entire capsule, as a whole, consists of (1) capsule membrane, (2) the inclosed protoplasm, and (3) nucleus, to which may be added such non-essential structures as vacuoles, pigment granules, crystals, &c. The nucleus behaves in every respect like a true cell-nucleus, and thus lies at the base of the universal opinion, that the whole organism, notwithstanding its variations, is unicellular, and remains through life a true individual cell.

The outer capsule (*extra-capsulum*) includes all those parts of the soft body which lie outside the central capsule. It consists of a gelatinous mantle, which completely surrounds the central capsule, but is separated from its outer surface by a thin layer of protoplasm. The pseudopodia radiate from this thin layer, penetrate the gelatinous mantle, and form a network on its surface, whence they extend into the surrounding water.

Cienkowski was the first to observe that the "yellow cells," already referred to as supposed "sarcoblasts," live independently after the death of the Radiolaria, and in consequence that they do not belong to it, but are unicellular parasitic algæ, to which the name of *Xanthellæ* has been given.

The pseudopodia, or thread-like processes, exhibit the same peculiarities as in all true Rhizopods ; they are usually very numerous, long and thin, flexible and sensitive filaments of sarcode, which exhibit the peculiar phenomena of granular movement. They serve as active organs for the inception of nutriment, for locomotion, sensation, and the formation of the skeleton.

The skeleton of the Radiolaria, says Haeckel, "is developed in such exceedingly manifold and various shapes, and exhibits at the same time such wonderful regularity and delicacy in its adjustments, that in both these respects this group excels all other classes of the organic world. For, in spite of the fact that the Radiolarian organism always remains merely a single cell, it shows the potentiality of the highest complexity to which the process of a skeleton formation can be brought by a single cell." The chemical composition of the skeleton, in many cases, is pure silica, or flint; in some it is a silicate of carbon, and, in a few, of a peculiar organic substance, called *acanthin*. Calcareous skeletons, or skeletons composed of lime, do not occur. In the great majority this skeleton has the form of a delicate lattice shell, or a receptacle, in which the central capsule is enclosed. In a small minority, however, this is not the case. The skeleton then consists only of isolated rigid pieces, or of a simple ring, or of a basal tripod, with or without a loose tissue, and the central capsule is not then surrounded by a special latticed receptacle, but only rests upon the skeleton.

We pass on, from this brief description, to what is known, up to the present, of the development of the Radiolaria, as set forth by Haeckel.[1]

It may be assumed that in all Radiolaria, on

---

[1] "Report on the Radiolaria of the *Challenger* Expedition," by Professor Ernst Haeckel, vol. xviii. (1887), p. xciii.

maturity, the central capsule discharges the functions of a sporangium, and its contents are broken up into numerous flagellate swarm-spores, or zoospores. After these have emerged from the ruptured capsule, they probably pass into an *actinophrys* stage, and then, after the formation of a jelly-veil into the condition of *sphærastrum*. Afterwards, when a membrane is formed between the outer jelly-veil and the inner nucleated cell-body, there is what has been termed the *actissa* stage, which exhibits in its simplest form the differentiation of the spherical central body into central capsule and outer capsule.

The zoospores generally arise in the following manner :—The nucleus of the central body, breaks up into numerous small nuclei, and each of these surrounds itself with a small portion of the protoplasm. From the resulting small, roundish, or ovoid cells protrudes one or more vibrating threads (vibratile flagella). The fully-developed spores, which commence their vibrations even within the central capsule, emerge when it ruptures, and swim about freely in the surrounding water by means of the thread or flagellum. Their form is for the most part ovoid, or rather cylindrical, sometimes spindle-shaped, and from 4 to 8 μ in diameter. Immediately behind the flagellum, at its base, lies a spherical nucleus, with several small, flat granules in the posterior portion. The number of vibratile flagella appears to be variable, usually one, but sometimes more.

The fate of the flagellate zoospores is uncertain, as all attempts to rear the swarming zoospores has failed, but, from analogy, the hypothesis seems to be fully justified, that they are converted into an intermediate *actinophrys* stage, in which the body becomes spherical, with fine pseudopodia protruding all round, whilst the nucleus assumes a central position.

What has been termed the *sphærastrum* stage is believed to succeed. This stage arises from the *actinophrys* stage by the excretion of an external jelly-veil. In this condition the young Radiolarian is a simple cell, with pseudopodia radiating on all sides, its body consisting of three concentric spheres, viz., the central nucleus, the protoplasmic body proper and the surrounding jelly-veil. When a firm membrane is developed between the last two spheres, this stage passes over to the next, or *actissa* stage. Probably in all Radiolaria the *sphærastrum* stage develops immediately into the typical *actissa* stage, by the formation of a firm membrane between the protoplasmic body and its jelly-veil. Thus arrives the simplest form of Radiolarian organisation, with the central body differentiated into an inner capsule and outer capsule, by means of the intervening membrane.

The general course of individual development begins with the formation of zoospores in the central capsule, but in some groups there is a different process introduced by the simple division of the unicellular organism. This spontaneous division is common in the social Radiolaria, and produces their colonies. In these cases the increase by division is nothing else than an ordinary case of cell-division, in which bisection of the nucleus precedes that of the central capsule. Another mode of growth in the colonies is the multiplication of the central capsules by gemmation. The gemmules or capsular buds are developed on the surface of the young central capsules before they have secreted a membrane. "They grow, usually in considerable numbers, from the surface of the central capsule, which is sometimes quite covered with them. Each bud usually contains a raspberry-like bunch of shining fatty globules, and, by means of reagents, a few larger, or a considerable number of smaller; nuclei may be recognised in them. The naked protoplasmic body of the bud is not enclosed by any membrane. As soon as the buds have reached a certain size, they are constricted off from the central capsule, and separated from it. Afterwards each bud becomes developed into a complete central capsule, by surrounding itself with a membrane, when it has attained a definite size.

A peculiar method of reproduction, which has been characterised as an "alternation of generations," occurs in the social Radiolaria, distinguished by the production of two different kinds of swarm-spores. These two kinds are called *isospores* and *anisospores*. The isospores correspond to the ordinary asexual zoospores already described, developing without copulation. The anisospores, on the other hand, are sexually differentiated into female zoospores (*gynospores*) and male zoospores (*androspores*). The female, or gynospores, are larger, less numerous, possess larger nuclei, and have a fine filiform network. The male, or androspores, are much smaller, more numerous, with smaller nuclei and thicker tubercles. In some cases both kinds are developed in the same individual, but not always. It is probable that these two forms of anisospores copulate with each other, after their exit from the central capsule, and thus produce a new cell by the simplest method of sexual reproduction. But since the same species which produce these sexual anisospores at other times give rise to the ordinary or asexual zoospores, it is possible that these two forms of reproduction alternate with each other, and that they thus pass through an alternation of generations.

The nutritive materials which these animals require for their support are derived partly from foreign organisms, which they capture and digest (although Dr. Wallich has declared his discredit of this), and partly from the unicellular algæ (*Xanthellæ*) which live within them. The considerable amount of starch, and starchy products, elaborated by these tenants (for they are not truly parasites), as well as their protoplasm and nucleus, are available, on their death, for the nutrition of the Radiolaria which harbour them. Nutrition, by means of other particles, obtained by the pseudopodia from the surrounding medium, is by no means excluded; indeed, it is certain that numerous Radiolaria are nourished for the most part, or wholly, by this means. Diatoms, infusoria, as well as decaying particles of animal and vegetable tissues, can be seized directly by the pseudopodia, and conveyed, either to the surface of the jelly-veil, or that of the central capsule, in order to undergo digestion there. The indigestible constituents are collected, often in large numbers, and removed by the streaming of the protoplasm. Professor Haeckel seems to be firm in his opinion on this point, for he alludes to those who have doubted it, and then affirms:—"I must, however, maintain my former opinion, which I have only modified in so much that I now regard the outer surface of the jelly-veil (*calymma*), rather than the outer surface of the central capsule, as the principal seat of true digestion and assimilation."

Those bodies called "yellow cells," which at one time were a mystery, and afterwards declared to be statoblasts, are now held to be yellow unicellular algæ, of the group *Xanthellæ*, living within the substance of the Radiolaria, and assisting in their support. The animal cells furnish the algæ with shelter and protection, and also with carbon dioxide, and other products of decomposition, for their nutriment, whilst, on the other hand, the vegetable cells of the *Xanthellæ* yield the Radiolarian its most important supply of nutriment, protoplasm, and starch, as well as oxygen for respiration. It has been experimentally proved that Radiolaria, which contain numerous *Xanthellæ*, can exist, without extraneous nutriment, for a long period in closed vessels of filtered sea-water, kept exposed to the sunlight, the two organisms furnishing each other mutually with nourishment. In many Radiolaria the algæ are entirely wanting, therefore they are not absolutely necessary for existence.

Circulation is the general term used to express certain slow currents in the protoplasm, within and without the central capsule, of these organisms. These currents probably continue through the whole life of the animals, and are of great importance for the performance of their vital functions. Sometimes the circulation is directly perceptible in the protoplasm itself, but it is usually only visible owing to the presence of suspended granules. Although the protoplasm of the inner capsule is in communication with that of the outer, through the openings in the capsular membrane, nevertheless the currents exhibit certain differences in the two portions. It is not so easy to observe the movements within the inner capsule as without it, but it is sometimes possible to observe the granules pass through the openings in the capsule membrane. A distinct flowing of the plasma outside the central capsule may be readily observed in all Radiolaria in the living state. This is most easily seen in the free pseudopodia, which radiate from the surface into the surrounding water. In general the direction of the streams is radial, and it is frequently possible to observe two streams, opposite in direction, the granules on one side of the radial thread moving outwards, whilst those on the other side move inwards. If the threads branch, and neighbouring ones become united by connecting threads, the circulation may proceed quite irregularly in the network thus formed. The rapidity of the currents is subject to considerable variation.

In addition to the interior movements of the plasmatic currents, there are also two groups of motor phenomena which may be observed, the contraction of individual parts, which produce modifications of form, and the voluntary or reflex motion of the whole body. Active locomotion, which is perhaps voluntary, occurs in three modes: (1) The vibratile movement of the flagellate swarm-spores, (2) the swimming of the floating organisms, (3) the slow creeping of those which rest accidentally upon the bottom. .

The movement of the swarm-spores by the oscillation of the thread, or flagellum, does not differ essentially from that of ordinary flagellate infusoria. The swimming of mature Radiolaria is confined to a vertical direction, causing the rising or sinking in the water. This is probably due to increase or diminution of the specific gravity, which is perhaps brought about by the retraction or protrusion of the pseudopodia. The most important organ, however, is probably the jelly-veil, by the contraction of which the specific gravity is increased, while it is diminished by its expansion. The slow creeping locomotion only occurs when the animal comes in contact with a solid surface, and is possibly due to muscle-like contractions of the pseudopodia.

On the subject of phosphorescence in the Radiolaria, Professor Haeckel remarks that "many Radiolarians shine in the dark, and their phosphorescence presents the same phenomena as that of other luminous marine organisms; it is increased by mechanical and chemical irritation, or renewed if already extinguished. The light is sometimes greenish, sometimes yellowish, and appears generally (if not always) to radiate from the fatty sphæres of the inner capsule. Thus these latter unite several functions, inasmuch as they serve, firstly, as reserve stores of nutriment, secondly, as hydrostatic apparatus, and thirdly, as luminous organs for the protection of the Radiolaria. The production of the light depends, probably, as in other phosphorescent organisms, upon the slow oxidation of the fat globules, which combine with active oxygen in the presence of alkalis."

As to sensation in the Radiolaria, it must be admitted to be in an inferior degree of development. For although subject to various stimuli, and possessing a certain power of discrimination, no special sensory organs are differentiated ; the pseudopodia acting both as organs of motion and sensation. In general they seem to be sensitive to pressure, to temperature, to light, and to the chemical composition of the sea-water. The reaction, subject to these stimuli, correspond to the sensation of pleasure, or dislike, which they elicit, in the motion of the protoplasm, changes in the currents, and in contraction of the central capsule and the pseudopodia. Professor Haeckel considers that the central capsule contains the common central vital principle, which he terms the "cell-soul," and that it may be regarded as a simple ganglion cell, comparable to the nervous centre of the higher animals, whilst the pseudopodia are analogous to a peripheral nervous system.

Only a few words are requisite on the distribution of these organisms, since they are found in seas all over the world, under all climatic conditions, and all depths. To say as much as this is to recognise them as universally scattered, not only in tropical regions, but to the North and South Poles, where their limit has not yet been found amongst the eternal ice, in which their skeletons are imbedded. The richest development of forms, and the greatest number of species, undoubtedly occur between the tropics, whilst the frigid zones exhibit great masses of individuals, in but comparatively few genera and species. "The surface of the open ocean seems everywhere, at a certain distance from the coast at least, to be peopled by crowds of living Radiolaria." Thomson and Murray were convinced, as one of the results of their experience, that "these animals exist at all depths of the ocean, and that there are large numbers of true deep-sea species which are never found on the surface or at slight depths." It has been admitted that the great majority of species, which have hitherto been observed, have been obtained from the bottom of the deep sea, and more than half of all the species have been derived from the pure Radiolarian ooze, which forms the bed of the Central Pacific, at depths of from 2,000 to 4,000 fathoms.

The Geological Distribution of Radiolaria is also remarkable, since they are found in all the more important groups of the sedimentary rocks of the earth's crust. The great majority belong to the Cainozoic or Tertiary period, and, in fact, to its middle portion, the Miocene period. To this period belong the "Polycystin Marl" of Barbados, the Nicobar clay, and the Mediterranean Tertiary deposits. From the Mesozoic, or Secondary period, many well-preserved species have been described. All the main divisions of the Jura, both the upper and the middle, and especially the lower, appear, in certain localities, to be very rich in well-preserved shells of fossil Polycystina. The number obtained from the Palæozoic or Primary formations is much less than from the other two. A few species are now known from various Palæozoic formations, and not only from the Permian and the Coal measures, but also from the older Devonian and Silurian systems. An important fact is recorded by Haeckel as to the connexion between these fossil and extant living forms. He says that "amongst the Miocene Radiolaria numerous species are not to be distinguished from the corresponding living forms. On the other hand, certain genera, which are rich both in species and individuals (recent as well as fossil), present continuous series of forms, which lead gradually and uninterruptedly, from old Tertiary species to others still living, which are specifically indistinguishable from them."

The skeletons of the Radiolaria are so beautiful, and of such a variety in modification of their form, that they are deservedly held in high esteem as objects for the microscope, even when not studied with a higher object. They are, however, of still higher interest when viewed as the homes, or skeletons, of the simple organisms that inhabit them. It would be hopeless to attempt to give any adequate idea of their beauty, or variability, by description, hence we shall rest content with some general remarks, as a conclusion to this chapter.

In the majority the skeleton is a delicate latticed shell, or receptacle, which encloses the central capsule. In a minority, it consists only of isolated rigid pieces, of a ring, or of a tripod. According to Haeckel there are twelve principal forms which may be regarded as morphological types of skeleton formation. There are, for instance, the Asteroid or star-shaped skeletons; the beloid or spicular, which consist of several disconnected portions; the lattice spheres; the lattice ellipsoids; the lattice disks; the larcoid lattice shells; the cyrtoid lattice shells; the circoid skeleton, with a simple vertical ring; the plectoid, in which three or four siliceous spines proceed from a common point; the sponge-like skeleton, in which a kind of wicker-work of a more or less spongy structure is developed; tubular skeletons; and conchoid skeletons of bivalved lattice shells.

The lattice structures are of extremely variable form, but the specific conformation depends mainly upon the radial spines. In some of these the entire lattice shell is excreted simultaneously. In others they are developed from separate lattice plates, deposited in succession.

"The skeleton in the great majority is armed with radial spines, which are of great importance in the development of their general form, and of their vital functions. The number, arrangement, and disposition of the spines is usually the determining factor in the general form of the skeleton. Physiologically they discharge distinct functions, as organs of protection and support; they act also, like the tentacles of the lower animals, as prehensile organs, since their points, lateral branches, barbed hooks, &c., serve to hold fast nutritive materials. In general, main spines and accessory spines may be distinguished: the former are of pre-eminent importance in determining the figure of the skeleton, the latter are merely appendicular organs."[1] Accessory spines may include all those processes which have no influence in the determination of the form of the skeleton as a whole. They are developed in the utmost variety, sometimes as hairs or bristles, sometimes as thorns or clubs, either straight or curved, smooth or barbed, sometimes standing vertically upon the shell, or directed towards the centre, sometimes obliquely, or rising at a definite angle.

---

[1] *Challenger:* " Report on the Radiolaria," p. lxxxix.

The most satisfactory initiation into the mysteries of construction, in these elaborate skeletons, can best be obtained by the study of the splendid atlas of figures of the Radiolaria, published in connexion with the series of *Challenger* reports.

CHAPTER IV.

SPONGE WEAVERS.

WHAT is a sponge? is a question much more readily asked than answered. Not so many years ago a speedy answer would have been given, even if not satisfactory. In those days there was a rough-and-ready classification of animals, plants, and minerals in vogue, subject to three primitive definitions,—animals possessing life and motion, plants life without motion, and minerals without life or motion. Between this triumvirate was the world divided, until it became manifest that there were undoubted animals which were as fixed as plants, and undeniable vegetables which were as free in their movements as animals. The old formulæ were no longer to be relied upon, and new definitions attempted, but with these we will not trouble; since all definitions are but temporary, and but relatively satisfactory. Even now there are some persons who require some persuasion before they will accept the free swimming diatoms as real vegetables, or the fixed and plant-like zoophytes, or more robust and fungus-like sponges, as animals. Whether any such persons ever experimented upon a piece of common bath sponge by burning it, and sniffing the odour we are unable to say; but we should doubt their judgment, or the acuteness of their sense of smell, if they had not at once referred the odour to something akin to burning hair rather than to charred wood.

We may start with the assumption that sponges are animal, and not vegetable, because, as we proceed, this assumption will be abundantly proved, although the animals may be lower in rank, and even more simple than the zoophytes, and their individuality less easily demonstrated. Because they flourish whilst attached to the rock, by means of a sort of disk at the base, in like manner to the seaweed, proves nothing, for all further resemblance is at an end, there being only an analogy and no affinity between them.[1] It is a common mistake,— into which even men with some claim to be considered scientific occasionally fall,—to confound analogy with affinity. Things may be very much like each other, in more particulars than one, and yet possess no relationship whatever. How many

---

[1] For an account of the History of Sponges, from the earliest times, see the introductory chapters to "A History of British Sponges and Lithophytes," by George Johnston, London.

fallacies in our own day flourish for awhile upon this fundamental error, but time and experience settles the question at last.

It is by no means astonishing that the majority of persons, not being naturalists, should at the mention of the name of a sponge at once associate it with those soft, flexible, and very useful little articles which accompany a bath, or dangled by a string to their slates in school-boy days. And certainly these are sponges of a certain kind; but the general idea of a "sponge" must include more than this. Not only must it include all the many varieties of sponges, —good, bad, and indifferent, which are employed for ablutionary purposes,—but also very many others which could not by any possibility be converted to such a use. Nearly every local museum exhibits, as one of its choice treasures, a gigantic "Neptune's goblet," in shape somewhat like a goblet or vase, perhaps as much as three feet high, and more than a foot in diameter, and this is also a sponge. Then, again, those elegant white, latticed, horn-shaped objects, called "Venus's Flower-basket" (fig. 15), which occupy a more favoured position in the local museum, or stand exposed for sale in the window of any "naturalist" of moderate pretensions, are also sponges, but *not* washing sponges. Indeed, there are sponges of all sizes, innumerable forms, and varied characteristics, which scientific people class

112 *TOILERS IN THE SEA.*

in three principal groups. First, there are the horny (or keratose) sponges, such as the washing sponge,

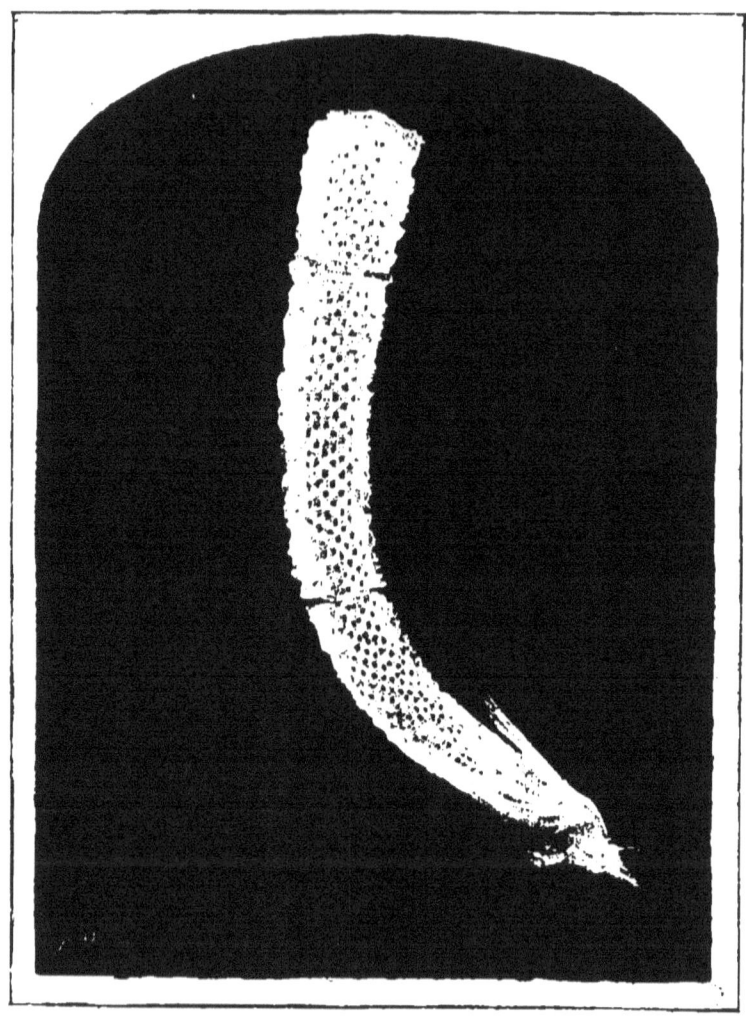

FIG. 15.—VENUS'S FLOWER-BASKET.—(*Euplectella aspergillum*, Owen.)

which are soft, flexible, and of a substance somewhat resembling horn, but flexible, and chemically different from horn; then, secondly, there are the flinty (or siliceous) sponges, to which Venus's Flower-basket belongs, in which the skeleton is rigid, composed of flinty fibres, not unlike "spun glass"; and lastly, the chalky (or calcareous) sponges, in which the skeleton is mostly chalky, or carbonate of lime. For general purposes this division is useful, although some exception might be taken to it for scientific purposes; nevertheless, they are all sponges, and the idea of a sponge should be so extended as to embrace them all. Most of them live and flourish in the sea, whilst only a very few occur in fresh water.

Accepting this general notion of a sponge, our best authority on the subject says:[1]—"Whatever may be their form, or however they may differ from each other in appearance, there are certain points in their organisation in which they all agree. In the first place, however variable in its form and mode of structure, there is always a skeleton present, in which the rest of the organic parts are based and maintained. Amidst this skeleton, and intimately incorporated with it, are the interstitial canals, usually of

---

[1] "A Monograph of the British Spongiadæ," by J. S. Bowerbank, LL.D. Ray Society. 1864, &c.

two series; the first appropriated to the inflowing (incurrent) streams of the surrounding water, and the second to the outflowing (excurrent) streams, which they conduct from the interior of the sponge to the oscula at its surface, through which they are discharged. Enveloping the entire mass of the sponge we find the dermal membrane, in which are situated the pores for inhalation and imbibition of nutriment, and the supply of the inflowing canals, and the oscula, through which the excrementitious matter, and the exhausted streams of water, are poured from the terminations of the outflowing canals. These parts are indispensably necessary, and are always present in a living sponge. The attachment of the sponge to the body to which they adhere during life is effected by a basal membrane, which presents a simple adhesive surface, following the sinuosities of the body on which it is based, entering into holes, or cracks, and filling them up, thus securing a firm hold of the mass on which they are fixed."

The size and external form of sponges vary considerably, even in the same species; in the majority there is such a delightful irregularity as to baffle description. What the extreme size may be it is very hard to conjecture, but Neptune's Cup is sometimes three feet high, and there are minute species not so large as a pea; between these two extremes there is every gradation. Some of the forms are certainly elegant, being branched and forked like a stag's horn, but these perhaps are rather exceptional species, since the majority present, especially in the living state, little that is attractive in appearance. Their beauties must rather be sought in their microscopical structure, and the elaborate interweaving of the skeleton. It is probably on account of their unattractive appearance that they have had so few students, except here and there a plodding enthusiast, like our own Dr. Bowerbank and Dr. Carter; in fact, at almost any time during the past century those who pursued the study of sponges in these islands could be counted on the fingers of one hand. Of course, there are difficulties to be overcome, and one of these is the observation of specimens in the living state, for which purpose dredging must be resorted to, and such physical exertion as hardly commends itself to the majority of modern naturalists.

In order to give some general idea of sponge structure we must commence with the skeleton, but, before doing so, it will be necessary to explain what is meant by sponge "spicules," which are the element of which the skeleton in most sponges is mainly composed (fig. 16). If a fragment of siliceous or flinty sponge is boiled in nitric acid, all the fleshy portion is destroyed, whilst the flinty, or siliceous, remains unchanged. When this sediment is washed and transferred to an ordinary slip of glass and viewed under the microscope, it will be seen to consist entirely of little transparent fragments of flint, resembling glass, of regular and symmetrical form, sometimes very beautiful, largely intermixed with

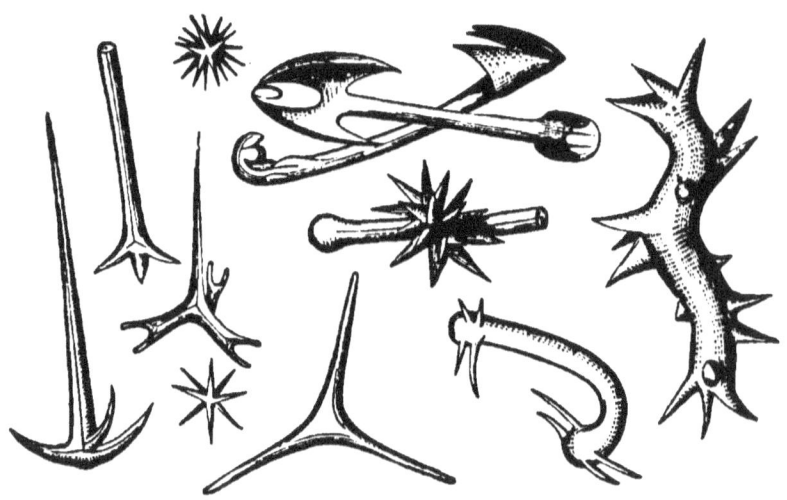

FIG. 16.—SPONGE SPICULES.

simple, straight, or curved bodies like needles and pins. These are the sponge spicules, each form having its own peculiar name, which is sometimes very peculiar and very composite, intended to indicate the particular form of "spicule." Dr. Bowerbank has enumerated somewhere about two hundred of these forms, but it would avail nothing to go over the catalogue of "Geniculated expando-ternate," "Unipocillated bihamate," "Tridentate equi-anchorate," or "Floricomo-hexradiate" spicules. Let it suffice, to indicate their great variety of form, referring to some figures in illustration (Plate III). Some of these forms occur chiefly in the principal skeleton, whilst others are confined to special parts, and have their own special function, chiefly of strengthening and supporting the sponge structure. It is noteworthy that these spicules being of flint, and the most indestructible portion of the sponge, are often found in most unexpected places, like the bones of a skeleton, when no other portion of the sponge is to be seen. They may be washed out from chalk, where they have lain buried for thousands of years, and even seen imbedded in a thin chipping of flint stone. The microscope has detected them mixed with sand, or associated with diatoms, in deposits laid down many generations ago, in the stomachs of molluscs, fishes, and even of marine birds of the present day; in fact, attached to almost everything that comes out of the sea, and in nearly every spot over which the ocean waters have ever flowed. It must not be forgotten that we allude in these remarks to the flinty spicules of the siliceous sponges, which are by far the most numerous, and not to the chalky or lime spicules of the calcareous sponges, which are not indestructible; the horny or keratose sponges being left out of the question, as the spicules are confined to a few species.

The skeleton of the horny or keratose sponges, such as the toilet sponge, is built up of horny fibres which branch and coalesce, and are interwoven into a kind of basket-work, sometimes strengthened by grains, sand, and other extraneous matter, or sometimes by a few spicules, but the most essential element is the horny fibre. In the calcareous sponges carbonate of lime, in the form of spicules combined with membrane, is the skeleton element; but in the siliceous or flinty sponges different forms and combinations of spicules cemented together form the skeleton (fig. 17).

"Sometimes the skeletons assume the shape of a beautiful regular or irregular reticulation, composed either of a nearly single series of elongate forms of spicules, cemented firmly together at their apices by keratode (which is the principal substance in the skeleton of the horny sponges), or by numerous spicules similarly cemented together, forming a strong and complicated fasciculated thread of reticulations. In other cases there is no reticulated structure, but the spicules are arranged in elongated compound bundles, which radiate from either the base or central axis of the sponge, whilst in others the reticulate and the radial system both enter into the structure of the skeleton, a modification of the network prevailing in the axis, and of the radial system towards the circumference of the sponge. Again, in other cases the spicules are simply and irregularly dispersed over the membranous base of the skeleton; and, finally, we find it simulating the form of pure keratose fibre, becoming a rigid and solid flinty fibrous skeleton."

Manifestly all this arrangement is destroyed either by the natural disintegration of the sponge, or the artificial breaking up by boiling in acid, as above stated, the cementing material being destroyed also, and nothing is left but a chaos of mixed spicules. The only method of observ-

FIG. 17.—SECTION OF SPONGE.

ing the arrangement is by cutting carefully very thin sections of the sponge, and submitting them to the microscope. Of whatever substance the skeleton is composed, and however it may be combined, it is an essential of all sponge structure, as much as the bony skeleton is of the structure of a mammal or a bird.

This skeleton must be covered with flesh, not of

the same kind as that of the higher animals, but of a peculiar half-transparent gelatinous substance, not altogether unlike isinglass jelly in appearance, and called *sarcode*. This substance corresponds to what we call flesh, but simpler in its organisation, and little more than a pellucid jelly; there is no other, or better, name which can be applied to it, since to call it *flesh* would be to confound it with something else, which it is not; but *sarcode* really means that it is "something like flesh;" therefore let it be sarcode. This substance is thinly spread all over the internal tissues, although not perfectly smooth, for sometimes it abounds in little obtuse elevations, separating occasionally into innumerable roundish or oval masses of variable size. Under a moderately high power of the microscope these masses are seen to contain granules of apparently extraneous matter in a shrivelled or collapsed condition. Occasionally there are found intermixed numerous flattened nucleated coloured cells, immersed in the sarcode.

Dr. Bowerbank says that "while the sponge, as a whole, is sensitive and amenable to disturbing causes, the sarcode does not appear to be especially so, as I have frequently observed a minute parasitical worm passing rapidly over the sarcode surface, and biting pieces out of its substance, without apparently creating the slightest sensation to the sarcode, or at all interfering with the general action of the internal organs of the sponge, and occasionally we find minute creatures permanently located in its large cavities, without appearing to cause it the slightest inconvenience."

At certain periods of the year portions of this sarcode, detached from the living animal, are capable of a certain amount of locomotion. It resembles, perhaps most nearly, the slow, gradual movements of that curious little creature, the Amœba, and consists in a constant change of form, with progress in different directions. Several independent observers have described these movements, and most of them have found them to occur at certain periods, and not at others, whatever the reason may be.

The living animal, or the units which go to make up the living animal of the sponge, has of late years been the subject of controversy, and there are still, perhaps, two principal opposed views,—one in favour of an analogy with the polypes, and the other with flagellate monads. The latter view is now altogether the most general. "The true essential part of a sponge," writes one who adheres to the latter view,[1] "is composed of structureless sarcode, and nucleated cells, placed side by side, with a flagellum, some cells having a hyaline collar protecting the flagellum. These latter cells line all the passages leading from

---

[1] "On the Natural History and Histology of Sponges," by B. W. Priest, in *Quekett Microscopical Journal*, vol. vi. p. 232. 1881.

the pores, in most cases to the cloacal cavity, or cavities, to the oscula, regulating the currents of water, and causing them to flow through the channels, and convey the nutriment necessary to the existence of the sponge. Some naturalists, I believe, look upon the collared cells as playing the part of respiratory organs only, and not as means for assimilating nutriment; at any rate, I have no doubt, along with others, with regard to their regulating the currents of water. In some species of sponges these ciliated cells occur only in well-determined circular chambers, with their ciliated ends pointing towards the centre, each chamber having a small aperture, which perforates the investing membrane. The late Professor James Clarke, of Kentucky, was the first to notice the analogy of these ciliated cells with the free flagellate collared infusoria, followed up at the present time by Mr. Saville Kent. It has been found by this gentleman that some of the free collared monads are identical with the ciliated collared monads discovered in the sponges, each separate collar-bearing cell possessing a separate existence, and securing its nutriment in the same manner. Furthermore,[1] Mr. Kent tells us that sponge structure may be, and *is*, built up from one of these constituent monads, by a

---

[1] "A Manual of the Infusoria," by W. Saville Kent, p. 143, and following. London, 1880.

repeated process of cleavage, by which means it quickly multiplies itself, though still more rapidly by the subsequent encystment and breaking up of the monads into spores."

It is quite unnecessary to recapitulate the varied phases of the controversy, which has become voluminous, on behalf of the theory advanced by Haeckel on the one hand and Professor Clarke on the other,[1] since this would scarce possess interest for the general reader; but should any one feel desirous of investigating the subject, he will find an admirable summary in the chapter, "On the Nature and Affinities of the Sponges," in Mr. Saville Kent's "Manual."

According to those who accept the affinity of the sponge animal to the flagellate infusoria, first distinctly propounded by Professor Clarke, and further developed since his death, the essential sponge-structure consists of three elements, namely, the collared flagellate monads, the hyaline mucous-like stratum (*cytoblastema*), and the amœboid bodies, or cells, to which is added the skeleton or framework already alluded to. It is claimed, on behalf of the flagellate monads, that they should hold the foremost position in the economy of the sponge, to which the mucous stratum and amœboid bodies are subsidiary.

---

[1] See "Nature of Sponges," by Henry Slack, in *Popular Science Review*, vol. xi. p. 167. 1872.

The essential feature of the flagellate monads is, that they possess a film-like collar of membrane, not unlike a little funnel, which is capable of extension or withdrawal, enclosing within it a terminal flagellum, or whip-like thread, and at the other extremity contractile vesicles (fig. 18). As to the collar, it is stated to be not a mere funnel-shaped expansion of inert sarcode, but a most active organ, having, during life and when fully extended, a continuous stream of fine granular protoplasm for ever flowing up the exterior, and down the interior, surface of the collar, identical with the cyclosis exhibited in the pseudopodia of Foraminifera. It has also been demonstrated that "this collar, with its characteristic currents, is an exquisitely contrived trap, for the arrest and capture of its customary food, which, driven by the action of the central flagellum against the outer margin of the collar, adheres to it, and passes, with the onflowing protoplasmic stream, into the animal's body." These collared monads are always found lining special cavities, excavated within the hyaline mucous-like stratum (or *cytoblastema*). There is, of course, in different sponges, variability in

FIG. 18. — COLLARED MONADS (*Halichondria panicea*).

the size and form of these cavities, and their precise location. In some they are distributed in a general manner throughout the internal canal-system ; whilst in others they are confined to sphæroidal chambers, excavated within the substance of the body of the sponge, but freely communicating with the inflowing and outflowing streams.

The second element is the common gelatinous base, or hyaline mucous-like stratum, in which the two other elements are imbedded. Although of the same consistence throughout, this stratum consists of two layers; that is, of an investing membrane in which no monad chambers exist, and a deeper, thicker substratum in which they are immersed.

The third element consists of the innumerable amœboid bodies, or cells, scattered more or less abundantly throughout the substance of the mucous-like stratum. These bodies have no distinct cell-wall, and unless specially sought after, are scarcely to be distinguished from the stratum in which they are imbedded. They vary much in outline, each furnished with a refractive nucleus, and are best seen in the investing membrane. " Like Amœba they are constantly undergoing a change of outline, and may also be observed to shift their position from one part to another of the matrix. Oftentimes their long, slender pseudopodia, radiating towards those of their neighbours, unite together, forming a complex network, which presents a remarkable resemblance to ganglionic corpuscles. It is undoubtedly through the stimulus received and transmitted by them that the characteristic contraction and expansion of the pores, or oscula, and other portions of the sponge body, are accomplished."[1] The relations between these amœboid bodies and the flagellate monads is at present little more than conjecture ; but Mr. Kent does not regard the former as independent structures, but rather as larval, or metamorphosed, phases of the collar-bearing monads, it having been clearly observed that the latter, when their course is run, lose their collar and flagellum, and become amœboid.

This endeavour to explain, as briefly as possible, the economy of what may be termed the sarcodous, or fleshy portion of sponge structure, has been necessary in order to exhibit the complexity of what at one time was considered a very simple matter, and dismissed with little more than an intimation that it was called " sarcode," and was the amorphous flesh of the sponge. Even now there is undoubtedly much more to be known, and further investigation, proceeding as rapidly as during the past few years, will elucidate that which is still dark and uncertain. If, in each individual sponge, we should come to

---

[1] Kent's " Manual of the Infusoria," p. 172.

recognise a colony of unnumbered workers, myriads of monads, living in company, toiling for the benefit of the commonwealth, and weaving a home for themselves in the great deep, enlarging, extending, and increasing the colony day by day, it will simply be a repetition, in another form, of the same story as told by the zoophytes, the sea-fans, the sea-mats, and the architects of the Coral Islands.

> "Each wrought alone, yet all together wrought
> Unconscious, not unworthy instruments,
> By which a hand invisible was rearing
> A new creation in the secret deep."

The entire sponge, in its living state, is enveloped by a sort of skin or dermal membrane, coated in the inside with sarcode, and strengthened in various ways by fibrous tissue, or spicules. This membrane has the power of opening and closing pores on any part of its surface, through which the animals breathe, or receive nutriment. Beneath these pores are large irregular cavities, which receive the water imbibed by the pores, and convey it by an inward current into the canals, which branch and ramify through all parts of the sponge, becoming smaller and smaller as they divide and recede. There is always a double series of canals, those which convey the water charged with nutriment, by an inflowing stream, to the remotest parts of the sponge, and those which collect the exhausted water, and transmit it back again by an outflowing stream, which find an exit in the oscula, or cloacal openings of the sponge. These latter are either dispersed or clustered together, according to the species, and are opened or closed at will. This we may term the circulatory system of the sponge structure, by means of which water charged with nutriment is inhaled, circulates, becomes exhausted, and is ultimately expelled, with the suspended rejectamenta. Dr. Bowerbank remarks that "the power of inhalation appears to be exerted in perfect accordance with the similar vital functions in the higher classes of animals, not involuntarily and continuously, as in the vegetable creation, but at intervals, and modified in the degree of its force by the instincts and necessities of the animal. And it may be readily seen that the faculty of inhalation is exercised in two distinct modes; one exceedingly vigorous, but of comparatively short duration, the other very gentle and persistent. In the exertion of the first mode of inhalation,—that is, during the feeding period,—a vast number of pores are opened, and if the water be charged with a small portion of finely-triturated indigo, or carmine, the molecules of pigment are seen, at some distance from the dermal membrane, at first slowly approaching it, and gradually increasing their pace, until at last they seem to rush hastily into the open pores in every direction. In the meanwhile the oscula are widely open, and pouring out with considerable force each its stream of the outflowing (excurrent) fluid (fig. 19); this vigorous action will sometimes be continued for several hours, and then either gently subside, or abruptly terminate. Occasionally a cessation of the action may be observed in some of the oscula, while in others it is proceeding in its full vigour, and sometimes it will be suddenly renewed, for a brief

FIG. 19.—PIECE OF SPONGE MAGNIFIED.

period, in those in which it had apparently ceased. These vacillations in the performance of its functions is always indicative of an approaching cessation of its vigorous action. When the vivid expulsion of the water has ceased, the aspect of the oscula undergoes a considerable change; some of the smaller ones gradually close entirely, while in the large

K

ones their diameters are reduced to half, or onethird, of what they were while in full action. Simultaneously with the decline in the force of the outflowing action, the greater portion of the pores are closed, a few only dispersed over the surface of the sponge remaining open, to enable the gentle inhalation of the fluid to be continued, which is necessary for the aëration of the breathing surfaces of the sponge. The breathing state of inhalation appears to be very persistent, and I have rarely failed in detecting it when I have let a drop of water, charged with molecules of indigo, quietly sink through the clear fluid immediately above an open oscule. These alternations of repose and action are not dependent on mere mechanical causes, and sponges in a state of quiescence may be readily stimulated to vigorous action, by placing them in fresh, cool sea-water, and especially if it be poured somewhat roughly into the pan, and agitated briskly for a short period ; and this will take place even in specimens that have very recently been in powerful action. No general law seems to guide the animal in the choice of its periods of action and repose, and no two sponges appear to coincide entirely in the time or mode of their actions. In fact, each appears to follow the promptings of its own instinct in the choice of its periods of feeding and repose."

We may infer that the chief source of nutriment

to the living sponge consists in the minute organisms, whether animal or vegetable, which are suspended in the water carried down the inflowing canals. A curious speculation has been indulged in by some authors, whether certain forms of the spicules, which enter into the composition of the flinty sponges, may not be instrumental in the capture and detention of little worms, and other animals, so that in fact they may be devoured for the sustenance of the sponge. One writer says : " I have for a long time entertained the idea that these elaborate and varied forms of defensive spicules, probably subserved other purposes than that of the protection of the digestive surface against the incursions of minute annelids, and other predacious creatures. They are admirably fitted to retain, and make prey of any such intruders. No small animal could become entangled in the sinuosities of the interstitial cavities of sponges, thus armed, without extreme injury from the numerous points of these spicules, and every contortion, arising from its struggles to escape from its painful and dangerous entanglement, would contribute to its destruction, and it may then, by its death and decomposition, eventually become as instrumental to the sustentation of the sponge as if actually swallowed by the animal."[1] Although there is little

---

[1] Bowerbank's " British Spongiadæ," vol. i. p. 122.

if any, direct fact in support of this supposition, it is at least a plausible one, and would serve to account for the presence of some forms of spicules, which seem to be scarcely intended solely for strengthening the fabric, or serving as a simple defence, but are evidently adapted for such a purpose as has been suggested.

Important as it must be to the sponge colony to protect itself, as much as possible, from its enemies, and also to ensure a supply of nutriment in its fixed position, it is of equal importance that provision should be made for the continuance of the species by the establishment of new colonies. For this purpose sponges are by no means left deficient in resources, and the result may be obtained by one of three methods, each of which can be resorted to in case of failure by the other. These three modes of reproduction are (1) by means of ova or eggs; (2) by gemmation, or budding; and (3) by the spontaneous division of the sarcode, or flesh. Taking these three modes in the order in which they stand, we commence with the reproduction by means of ova, or eggs, as at once the most common and universal in the animal kingdom. Here, again, there are two modifications, for although the production of eggs within an ovarium prevails in the majority of cases, yet it is possible for eggs to be engendered without a proper ovarium, although this method is still but imperfectly known. In the ordinary sponges of commerce, as dredged, and before they are cleaned and manipulated by the dealers, the microscope reveals myriads of minute ovoid bodies surrounding the fibres, immersed in the sarcode, which are believed to be the eggs, or ova, of this particular class of sponges, and are as yet the only kind of reproductive bodies found in them. In size and appearance these ova have a great resemblance to the ova produced in the more legitimate manner, and are about two micro-millimetres in diameter, so that it would take something like twelve thousand, placed side by side, to form a line an inch in length. This is almost all that is at present known about them, but they, or their analogues, do not appear to be confined to the keratose or horny sponges.[1]

---

[1] Saville Kent investigated this subject, in connexion with his "History of the Infusoria," and found them by thousands in several species of sponge. His conclusion is "that the collared sponge monad (already described), after assuming a quiescent state, divides by segmentation into a mass of characteristic microspores, and, these falling asunder, become distributed throughout the hyaline stratum." And again, "these spores distributed broadcast through the substance of the sarcode stratum (*cytoblastema*) may be met with and traced onwards through every intermediate size and stage, from the single sphæroidal spore, up to the adult collared monads, or amœbiform bodies; the derivation of these spores through the splitting up into a granular or sporular mass of the entire

The ovaria, or egg-chambers, are spherical bodies of variable diameter, perhaps about one-eightieth of an inch in diameter, with a circular hole or opening at one end. The walls of this chamber are strengthened by spicules imbedded in the substance, which latter is of a like character to the other fleshy portions of the sponge. The interior of the chamber is lined with a delicate membrane, which encloses, and protects, the ova when they are developed. When mature the ova are extruded from the round hole,—the orifice of the ovarium.

The second mode of reproduction is by gemmation or budding, which may either be internal or external. The process of internal budding, as described by Dr. Grant,[1] appeared as "opaque yellow spots, visible to the naked eye, and without any definite form, size, or distribution, excepting that they are most abundant in the deeper parts of the sponge, and are seldom observable on the surface. They have no cell or capsule, and appear to enlarge by the mere juxtaposition of the monad-like bodies around them. As they enlarge in size they become

---

substance of the matured collar-bearing zooids being correspondingly demonstrated. In the sponge, all these transformations and developmental processes take place within the substance, which constitutes a suitable nidus for them."—*Infusoria*, vol. i. p. 174 and p. 176.

[1] *Edinburgh New Philosophical Journal*, vol. i. p. 16.

oval-shaped, and at length, in their mature state, they acquire a regular ovate form." When fully developed they separate from the parent, and pass out with the outflowing current at the oscules. At this period, Dr. Grant states, that they are endowed with spontaneous motion, in consequence of their larger extremity being furnished abundantly with cilia, or very minute transparent filaments, broadest at their base, and tapering to invisible points at their free extremities. After floating freely for a time, they become attached to some fixed body, adhering firmly to it, and spreading themselves out into a thin transparent film. When two of these come into contact they unite, thicken, and produce spicules, and in a few days there remains no line of distinction between them, and they continue to grow as one individual. Subsequently Dr. Bowerbank confirmed these particulars, except as regards the cilia in motion, and observes:—" That a sponge is not always developed from a single egg or bud, but that, on the contrary, many eggs or buds are often concerned in the production of one large individual, a few days probably serving, by this mode of simultaneous development, to form the basal membrane of the sponge, of considerable magnitude, as compared with the individual egg or bud, or with a sponge developed from a single egg only." As to his supposition that some of the buds are male, and some female, no corroboration has been obtained.

The third mode of reproduction, by the spontaneous division of the sarcode, or flesh, has long been known, and Dr. Bowerbank has remarked that "the truth appears simply to be that any minute mass of sarcode, whether separated voluntarily or involuntarily, has inherent life and locomotive power, and is capable of ultimately developing into a perfect sponge : and in the course of this process the dermal membrane is produced at a very early period." And afterwards he adds :—" Thus every description by these close and accurate observers tend to the conclusion that the multiplication of the sponge is effected by the origination in the egg, or by the agglomeration, in the form of buds, of particles of sarcode. The action of the minute masses of sarcode liberated by the bursting of the envelope of the egg, and their subsequent development, is precisely that of the so-called sponge cell liberated from the mass of sarcode lining the interstices of the sponge, and of the gemmules described by Grant, when sessile, each moves independently at first, each unites with its congeners into one body, and the results, both in means and end, are precisely the same, but their origin is different. The one is a generation of sarcode, within a proper membrane, in the form of an egg, while the others are the production of a bud (gemmule) by independent growth, or by spontaneous division of the sarcodous substance of the sponge."

Before leaving this portion of the subject, some brief allusion must be made to those bodies which have been denominated " ciliated sponge gemmules " or "ciliated larvæ," developmental forms of the " collared ciliated monads," referred to in describing the sarcode or flesh of the sponge structure. Mr. Saville Kent, who is responsible, not only for the name but the character of these forms, says that " the initial condition of these reproductive structures, as conceded unanimously by the independent testimony of every investigator, takes the form of an amœbiform body, varying in size from the three-thousandth to the two-hundredth part of an inch, and presents a considerable likeness to the primary condition of an ordinary ovum, or egg." " These amœboid oviform bodies are not independent products of the adult sponge stock, but simply retro-morphosed collar-bearing zooids that have retreated within the mucous stratum (*cytoblastema*), and assumed, as is common to them, after passing their matured collar-bearing stage, an amœboid condition. It is invariably found that these bodies are produced first in the deeper and consequently older portion of the sponge stock. This ciliated reproductive structure is in no sense an egg, or its derivative, but represents a coherent aggregate of monadiform swarm-spores, or, as it may be most appropriately denominated, a 'swarm-gemmule.' In their most characteristic form these reproductive bodies, or cell aggregates, consist of a uniform series of collared zooids." Forms are minutely described under about three types, as encountered in different species of sponge, but they agree in all essentials, as an aggregation of flagellate collared zooids, following the pattern of their parent.

Having now, somewhat imperfectly, sketched the outline of the structure, and economy, of a rather obscure but very large section of "Toilers in the Sea," we cannot dismiss them, humble as they are in their sphere and vocation, without calling attention to some of their beauties, which must appeal even to those who are wearied by details of their growth and development. It has been shown that there is in all a fundamental skeleton, of a horny, flinty, or chalky substance, held together and surrounded by the living flesh, or sarcode, of the animal, that this skeleton is secreted by the vital elements from the sea in which it dwells, that the sponge continues to grow, by aid of the sustenance it obtains from the water, that it contains ample provision in various ways for the reproduction of the species, and that some portions of the structures so elaborated are practically indestructible, and may survive the vicissitudes of thousands of years. The most beautiful example of a sponge skeleton with which we are acquainted are those elegant forms which, at one time so rare, are now comparatively common, called "Venus's Flower-basket," which is the bare skeleton of a flinty sponge. The ordinary size of one of these objects is from nine to ten inches in length, and about one and a-half inches in diameter at the top, curved and attenuated downwards, but with the upper two-thirds straight and erect, resembling in shape the figures of a cornucopia. The entire fabric is cylindrical, and hollow, looking like a delicate fairy-like basket-work, formed of elongated fibres, which consist of bundles of long thread-like spicules, crossed by similar bundles of spicules at right angles, so as to leave nearly square meshes, and these are crossed again at the angles, leaving round openings, resembling the cane-work of a cane-seated chair, but much finer, smaller, and more delicate. The top is covered with a similar network lid, composed of shorter spicules, and the attenuated base is surrounded by a beard of long filaments, like threads of glass, which have recurved hooks towards the ends. Throughout its entire length this fairy basket-work is ornamented still further with oblique concentric ridges, or furbelows of still more' delicate network, and a collar of like material forms a fringe round the lid. As seen in museums, and in the windows of naturalists, the whole structure is of a virginal whiteness, being nothing but the skeleton of what it was, but originally this entire skeleton was undoubtedly covered by the sarcode or flesh of the sponge. One of the earliest specimens brought to this country was sold for thirty pounds. Ultimately, more specimens were imported, which realised from ten to fifteen pounds each. Then, with fresh arrivals, the price sank from seven to three pounds a-piece, and, in the course of time, to ten shillings and less. Many of the specimens have a small crab in their interior, which it is suspected is often, if not always, introduced surreptitiously. When first known, it was currently believed that the case was spun by the crab, but this was among the natives of the Philippine Islands, those Europeans who believed in the legend were few in number.

As already intimated, these beautiful natural productions, which Dr. Gray observes [1] "will always be a most beautiful object, and an ornament to any room that may contain them," are sponge skeletons composed entirely of flinty spicules, of various shapes and sizes, interwoven like the most delicate lacework.

---

[1] "Venus's Flower-basket," by Dr. J. E. Gray, in *Popular Science Review*, vol. vi. p. 239. 1867.

It has been objected that it is an error to speak or write of these skeletons as interwoven, for there has been no interweaving, and no basket-work, since the material has been produced, and formed, at the same time as the framework, but no other word so well expresses the general appearance of this natural production as one borrowed from a branch of human industry, and it is scarcely probable that it will convey an erroneous impression.

Large and appreciable as most sponges are to the naked eye, it is to the microscope that we are indebted for a revelation of their beauties, for, except in a few rare and notable instances, their external appearance is by no means specially attractive. Whether we study the singular and variable forms presented by the spicules themselves, or their combination and arrangement in the sponge skeleton, they alike afford instruction and entertainment ; in the former case by their great variety and elegance, and in the latter by their adaptability to the special purposes which they are intended to serve. The answer to the question why any given form of spicule should have that particular form, in preference to any other, can only be answered by seeing the spicule in its natural position, and ascertaining what function it had to perform.

The most common and numerous forms of spicule are the simple cylindrical, or needle-shaped, if we may apply that term to them generally, whereas some are straight and others curved, some with a head like a pin, and others sharp at both ends, some rough and others smooth. Then again there are rayed spicules, some with three rays, some with four rays, like a cross, and some with many rays. These are sharp at the ends, or blunt, or spiny. Some again are shaped like hooks, or forks, or tridents, simple or complex. Others again resemble anchors, or wheels, or stars, or spiny spheres; in fact, to enumerate all the forms would be to produce an uninteresting catalogue, which a few figures would perhaps obviate. Suffice it to say that almost any typical form has numerous modifications.

These flinty spicules, which are peculiar to sponges, are practically indestructible (*see* fig. 16), and, from the position in which they are found, it is indisputable that sponges must have flourished in the ocean very many thousands of years ago. Dr. Bowerbank was of opinion that almost every flint stone was once a sponge, or had a sponge for its nidus. A writer on the " Spongeous Origin of Flints "[1] says, " If a thin chip or section of flint is submitted to microscopic examinations, sponge spicules in more or less

---

[1] "On the Spongeous Origin of Flints," by F. Kitton, in *Transactions of Norfolk and Norwich Naturalists' Society*, (1871-2), p. 51.

abundance, will invariably be seen." And again :—
"In the greensand large silicious nodules, known as Polypothecea, are of frequent occurrence, and when thin sections are examined their spongeous origin is distinctly seen ; these nodules were, however, formed under somewhat different conditions to the ordinary chalk flint, the silica is distinctly crystalline and doubly refractive, and polarises like quartz or agate ; the sponges were also probably different from those belonging to the chalk. A careful microscopic examination of very many sections did not reveal the presence of any form of spicule, they were most likely allied to the recent keratode sponges ; in fact, a thin slice of ordinary domestic sponge greatly resembles a section of its silicified predecessor. The reticulations are not solid, but tubular, and I have been able, in many cases, to fill them with colouring matter."

No one ventures to doubt, or dispute, the evidence, that sponges are amongst some of the oldest inhabitants of this world of ours, and on that account, if on no other, are worthy of our better acquaintance. We have endeavoured to compress, within a few pages, an epitome of a history which would readily have expanded sufficiently to have filled a volume, but even this fragmentary sketch will show, that there is something worth learning in the "home without hands," constructed in the sea depths, by the kindred of that lump of "common sponge" which has helped us so often in our ablutions. Beyond this, it may even lead to a kindlier affinity for the sea and its wonders, and to join with Kingsley in singing :—

> "Thou sea, who wast to me a prophet deep
> Through all thy restless waves, and wasting shores,
> Of silent labour, and eternal change ;
> First teacher of the dense immensity
> Of ever-stirring life, in thy strange forms
> Of fish, and shell, and worm, and oozy weed :
> To me alike thy frenzy and thy sleep
> Have been a deep and breathless joy."

CHAPTER V.

PLANT-ANIMALS, OR ZOOPHYTES.

AN enthusiastic naturalist, who wrote and published most interesting records of his "Rambles," has given the following note of one of his experiences. "Every morning," he says, "when the weather was good, one of us went out in a boat." This was on the Sicilian coast, where he saw the sea under an aspect entirely new to him, in an absolute and profound calm, with the surface as smooth as a mirror, permitting the eye to distinguish the minutest details at an incredible depth. "Leaning over the side of the boat, we could see, flitting beneath our eyes, a vision of plains, valleys, and hills, in one place with bare and rugged sides, in another, clothed with verdant herbage, or dotted over with tufts of brownish shrubs, and in all respects calling to mind the distant view of a passing landscape. But it was not the varied outlines of a terrestrial scene on which our eyes were riveted, for we were scanning the rugged contour of rocks, more than a hundred feet below us, amid submarine precipices, along which the undulating sands, the sharply cut angles of the stone, and the rich tufts of brightly-coloured red weeds, and glossy fucus fronds, lay revealed to sight with such incredible preciseness and clearness, as completely to deprive us of the power of separating the real from the ideal. After gazing intently for a while at the picturesque scene beneath our eyes, we scarcely perceived the intervening liquid element which served for its atmosphere, and bore us on its clear surface. We seemed to be suspended in empty space, or, rather, realising one of those dreams in which the imagination often indulges, we appeared to be soaring like a bird, and to contemplate from some aërial height the thousand varied features of hill and dale." Amongst the things to be seen were "a hundred different kinds of Polyzoaries, blooming in tufts of living flowers, or ramified into little shrubs, every spur and bud of which was an animal, and which, by their interlacing stems, their variegated branches and budding shoots, can scarcely be distinguished from true vegetables." [1]

Few of us have been privileged to see these "Plant-animals" flourishing in their native element,

---

[1] "Rambles of a Naturalist," &c., by A. de Quatrefages. (English edition.) Vol. i. p. 175. London, 1857.

but all of us, perhaps, at some time or other, have made the acquaintance of their skeletons, or, more accurately, their empty and deserted houses, torn from the rocks by the storm, and cast upon the beach

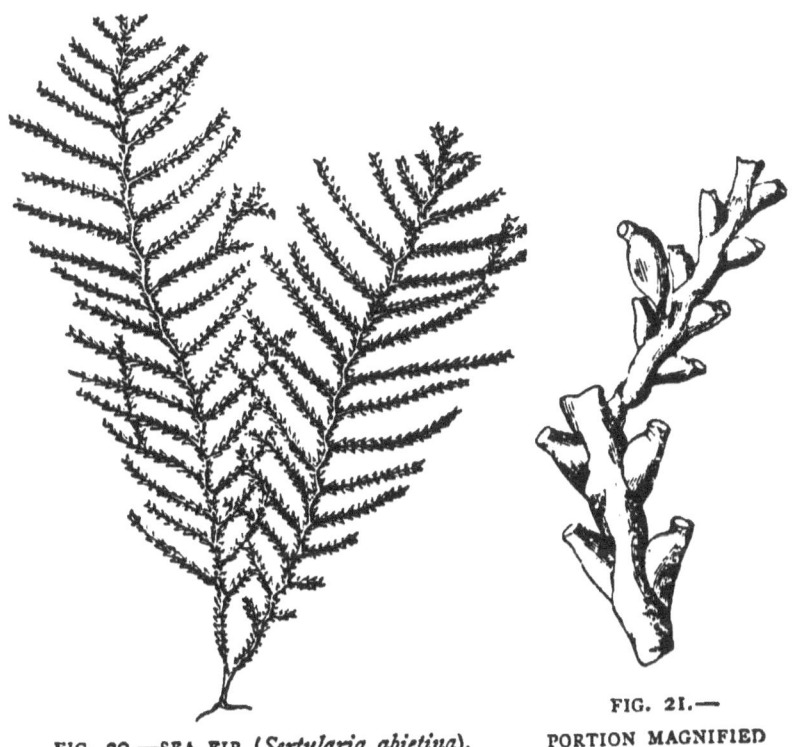

FIG. 20.—SEA FIR (*Sertularia abietina*). FIG. 21.— PORTION MAGNIFIED

(figs. 20, 21,). Tiny branched, horny, but extremely delicate twigs, by some called "Corallines," but improperly, and by others "Zoophytes," common enough on every shore, especially after rough weather.

"Large tangled masses of them, which are full of beauty in themselves, are cast ashore, and if examined while still fresh, and moist, will often be found to conceal some of the smaller kinds in a living state." It would be an unwarrantable presumption on our part to suppose any stray visitor to the sea coast ignorant of such pretty objects, and withal such "common objects of the sea shore", as Zoophytes, which may be seen adhering to every scallop-shell in the town fishmonger's window. Some may even have admired them when cleaned and grouped in fanciful shapes, for ornamental purposes, as offered for sale in the shop windows at fashionable watering-places.

Should any mind be still harassed by doubt, a glance at one or two of the woodcuts may serve to restore peace. It is hardly surprising that, in days gone by, these organisms were regarded as sea-plants, kind of sea-weeds. Fixed to the rock, branched and divided (fig. 20), and bearing capsules, resembling a kind of fruit, so long as their minute structure, and method of life, were unknown, it was excusable that such mysterious "Toilers in the Sea" should be deemed vegetables. Now, when their innermost secrets have been revealed, no one would hesitate to pronounce them animals, colonies of little animals, building for themselves these horny, plant-like homes, and dwelling in them.

That portion of the colony which is persistent and permanent, is the branched and horny *polypary*;

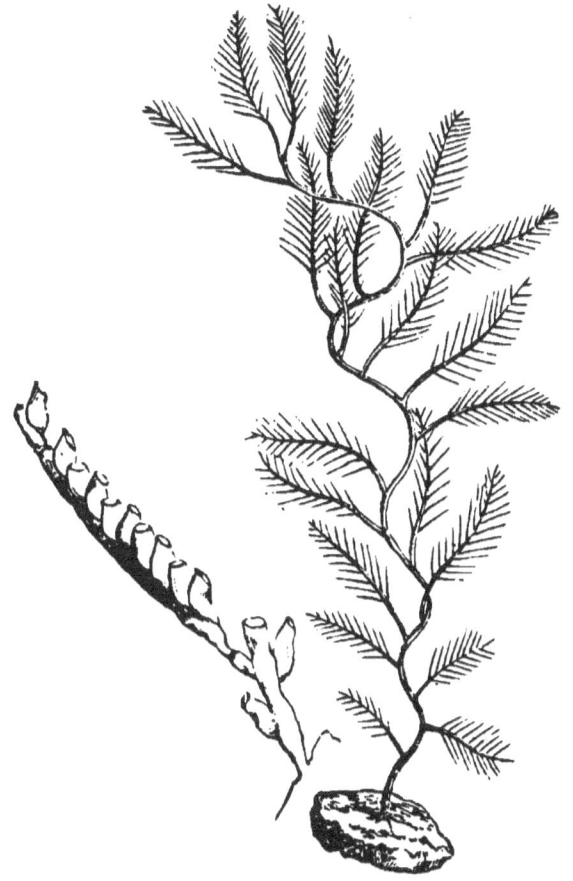

FIG. 22.—SICKLE CORALLINE.

this is the part cast on shore after a storm. The living animals are contained within these horny branched tubes, which are slender, and hollow, and very elastic, with joints at short and regular intervals. The little animals soon die, when removed from their native element, and shrivel within the tubes. For the most part the 'polyparies, or homes of the polyps, are of a yellowish or horny colour, and when immersed in water recover speedily their original form, and elasticity. The material of which this sheath is formed is called chitine, which resembles horn in many particulars, but differs from it in its chemical composition. It would be more accurate therefore to describe these polyparies as branched, chitinous, amber-coloured tubes, which shelter and protect the polyps, or little animals, which inhabit them.

There are many forms, or species, of these zoophytes, differing in the form of the polypary, or in some minute detail, of the animals which dwell in them. Some kinds prefer shallow, others deep water. Some attach themselves to the rocks, others to shells, or the larger seaweeds, and again some of the smaller are parasitic upon the larger species. Each kind, or species, has its own peculiar habit, which, for our present purpose we need not particularise. In some species the polypary is solitary, or scattered singly over the rock, stone, or shell, whilst in others it forms dense tufts; but these are minute details which do not influence the general observations it is our province to offer; this object being to present, with as little technicality as possible, some general idea of the zoophyte animals, and the homes they inhabit. · We may premise that the animals themselves follow closely the type of the common Fresh-water Hydra, and hence, by studying either the red or green hydra of our "Ponds and Ditches," we shall

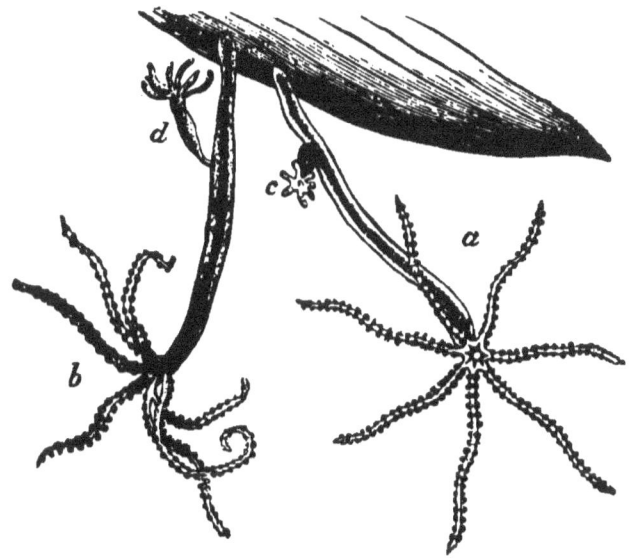

FIG. 23.—FRESH-WATER HYDRA (*a b* adults, *c d* young).

obtain a very fair idea of the hydra-shaped (hydroid) polypes of the marine zoophytes. There is, however, one important difference, which must be noted at once, in order to prevent any confusion of ideas. The Fresh-water Hydra (fig. 23), is solitary, not growing in colonies, like its marine cousins, although three or four young individuals are often met with on the body of their parent; and further, they are naked, not being enclosed in a chitinous covering, as is the case with the marine zoophytes.

These facts will be borne in mind, that the animals we are about to allude to, resemble miniature freshwater hydras, enclosed in a tube, and united together, by thread-like prolongations of the body, into compound branched colonies, of some hundreds of individuals, located, and enclosed within a common home, in some sense like all the lodgers in a model lodging-house, but with this difference, that there is a connecting link which unites all of them, and hence they are not entirely independent the one of the other. Divested of the chitinous shell, the whole colony would possess nearly the same form as that presented by the empty "skeletons" collected on the beach; the ultimate terminations of the branches, and branchlets, being occupied by the miniature hydra, which protrudes itself from the open end of the branchlet, and waves its tiny tentacles in the water, in search of food. It is a beautiful sight to gaze down into a clear rock pool, and observe one of these branched polyparies, with all its inhabitants out of doors, (fig. 24) portions of the body being thrust through the openings, at all the terminations, and the tiny tentacles waving like a star, from each individual, like flowers on an animated plant. At the least alarm the tentacles, and partly protruded body, are withdrawn into the openings of the polypary, and a minute trap-door closes over them, to protect them from injury. This is a remarkable contrivance, which, although not universal, is nevertheless common, each opening being furnished with a tiny operculum, or lid (fig. 25), fixed by a hinge on one side, which is pushed up when the animal protrudes itself, and falls back into its place when the retreat takes place.[1] It is no easy

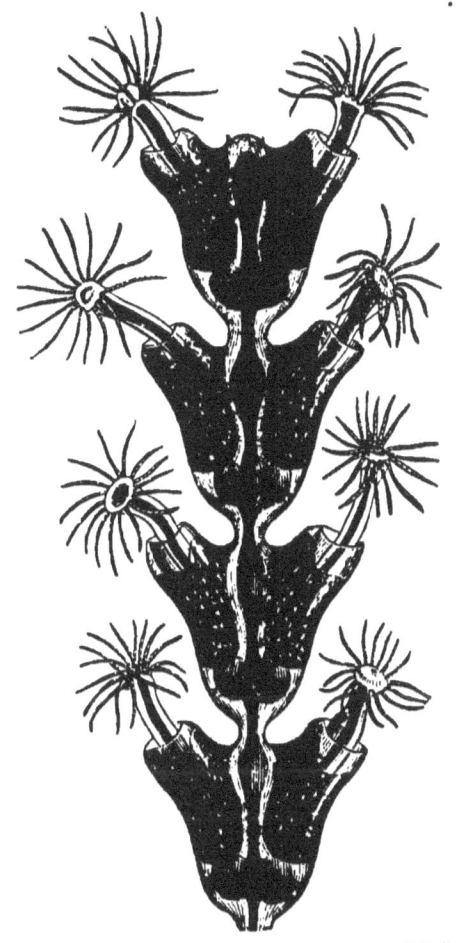

FIG. 24.—SERTULARIA PUMILA (*magnified*).

---

[1] Further information for those desirous of pursuing the

matter to obtain a good view of a colony with the animals extruded, as they retract themselves at the slightest alarm, and, when transferred from the sea to an artificial aquarium, there are few of them which will thrive, or, if they continue to exist, will manifest that vigour which characterises them in their native element. Many efforts have been made, with more or less of success, to kill the animals suddenly, when expanded so as to mount them in fluid, as permanent objects for the microscope. The theory is that if fresh water can be ejected upon them, when extruded, it kills them instantaneously, before they are able to withdraw; but it is not so easy to put the theory into practice, because a very slight motion of the water is sufficient to cause their withdrawal. Some of our friends claim to have achieved a fair amount of success, by carefully taking the living animals to a short distance only, in a vessel of sea-water, and then transferring them to a saucer, at the

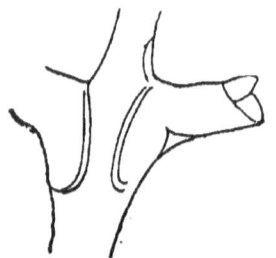

FIG. 25.—OPERCULUM OF SERTULARIA.

---

study of these animals, may be found in "A History of the British Zoophytes," by Dr. George Johnston (1847), or in the more recent "History of British Hydroid Zoophytes," by Thomas Hincks. London, 1868. Both published by Van Voorst.

same time keeping them immersed, placing them in the light, and maintaining the saucer perfectly still, till the expansion takes place, then instantaneously ejecting fresh water from a syringe upon them. Another mode has been suggested of putting the zoophytes in a small saucer of sea-water, and floating this on the surface of a large bowl of fresh water. As soon as a state of expansion which seems satisfactory is obtained, to swamp the saucer suddenly in the bowl of fresh water, by thrusting it down in an instant; but all these experiments require a quick and dexterous hand, so that failure is far more common than success. On the other side it may be urged that it has been accomplished, and may be accomplished again.

The method more usually adopted, and more strongly recommended by practical men,[1] is the substitution of spirits for water, say, gin or whisky, or spirits of wine. When the tentacles are fully expanded, the spirit is dropped upon the exposed animals, by means of a pipette. We are assured that this method can scarcely fail, with ordinary care, being far more certain than when fresh water alone is used.

To return from this digression to the animals in

---

[1] It is the method in ordinary use amongst the members of the Quekett Microscopical Club, and has their commendation.

question, we may observe that the polyps, which are attached to the common body, are those which suck in the nutriment for the support of that body, whilst the reproductive animals, or those which are concerned in the perpetuation of the species, soon become free, quit their old home, and go forth to establish a new one, being in themselves the unit from whence a new colony is to spring. For the present we are concerned only with the fixed polypes, the members of a given colony, who toil for the sustenance and support of the commonwealth. This, which is the true polype, consists of a digestive sac, with a terminal mouth, surrounded by tentacles, and is attached, at the base, to the extremity of one of the branchlets of the common trunk (*cænosarc*). The soft body is very contractile, and consequently variable in form, attenuated when thrust out at the openings of the horny sheath, and broader when withdrawn and contracted. The hollow cavity in the centre is a sort of simple stomach, ending upwards in an opening, or mouth, and continued below into the common canal, which traverses all the branches, down to the main trunk. The mouth is sometimes a simple, and sometimes a lobed opening, surmounting a sort of proboscis, which is either conical or funnel-shaped. It is an admirable little instrument for the acquisition of food, the lips opening and closing as occasion requires. The tentacles are arranged round the mouth, in one or two series; they are thread-like appendages, very flexible, and extensile, and are rough on the surface with a vast number of thread-cells, containing the only organs of offence and defence which these animals possess. These cells are imbedded in the flesh, and each contains a long delicate thread, or dart, often covered with barbs, and when at rest, lies spirally coiled up in the thread-cell. Undoubtedly these darts are deadly instruments when brought into operation, as they bury themselves in any soft substance with which they come in contact, and probably convey some poisonous fluid into the wound.

When the food is collected at the mouth, by means of the tentacles, and conveyed into the stomach, it is digested, and prepared for the support of the whole structure, and then passes down by the posterior opening into the canal, which traverses the whole colony. The circulation through this canal is very simple, the walls of the canal are lined with vibratile cilia, which, by their motion, keep up a continuous stream, from the stomachs of the polyps towards the trunk, and back again to the remotest branchlet, and thus every part of the structure is in communication with the myriad mouths, and is constantly supplied with all that is necessary to maintain the life, health, and vigour of the entire community.

The individual polypes often die, or become absorbed, whilst the colony, as a whole, still maintains its full vitality. In some species there seems to be a sort of winter rest, when numbers of the polypes are absorbed, but with returning spring, are again reproduced as plentifully as ever.

The growth and extension of the colony, or polypary, proceeds in a similar manner to the growth of a plant, by the production of buds. Every colony has originated from a single polype, and the plant-like form has been acquired by a succession of buds. "The Hydroid colony is enlarged by a purely vegetative process; fresh polypites bud rapidly from the prolific pulp, and in the larger and arborescent kinds, branches and branchlets germinate, according to the pattern of the species, from the zoophyte, as from the tree. For the multiplication and diffusion of the species, special zooids are set apart; and to them is allotted the reproductive function, as to the polype that of nutrition. They correspond with the flower-buds of the plant."[1]

> "Each wrought alone, yet all together wrought,
> Unconscious, not unworthy, instruments,
> By which a hand invisible was rearing
> A new creation in the secret deep."

Sexuality, in connexion with these zoophytes, has not as yet been alluded to, because the fixed polypes,

---

[1] Hincks's "Hydroid Zoophytes," p. xx.

to which our attention has hitherto been confined, are neuters, they are the workers of the colony, and, in some sense, correspond to the "workers" in a hive of bees. Now that we are required to describe the method in which new colonies are formed, and account for the perpetuation of the species, we must discover other members of the colony than those fixed, or permanent individuals, which furnish the colony with material support. The polypes are produced only at certain seasons, and occupy different positions in different species. They are produced in buds of a peculiar structure, borne upon an offshoot from one of the branches, and enclosed in a sort of urn, or receptacle, formed from the same chitinous, or horny, material as the covering, or skeleton of the colony. There may be some species in which this method does not prevail, in all its particulars, but in the most common species it is sufficiently general for our purpose.

The sexual bud (*gonophore*) consists of an outer coat, which serves for protection, and an enclosed, either male or female, zooid. Sometimes male and female occur on the same shoots, but in some cases they are produced on different colonies of the same species. The sexual zooids, which are in time developed from the special sexual buds, are either fixed to the colony, as the working polypes are, or they become free, and float away from the parent colony, and enter upon an independent existence. In the latter case they present such a contrast to the original stock, both in form and habit of life, that it seems difficult to believe that the new forms have any relation whatever to the old, but that they are rather new creatures, belonging to quite a different tribe. It is only by tracing their relationship that one is thoroughly convinced of the reality of the metamorphosis. In most cases the free zooid, which is the product of the sexual bud, assumes the form known to marine zoologists as that of the Medusa. For the benefit of those who are not marine zoologists we must explain what these Medusa-forms are like (fig. 26).

A small delicate crystal dome, a little more than a hemisphere, floating with the flat side downwards, and propelled through the water by contraction of the sides, its outline so delicate, and its substance so glassy, that it can be just distinguished in the water. The lower flat surface covered like a kettle-drum, which it resembles in miniature, with a thin membrane, with a round hole or opening in the centre. The margin of the flat surface apparently notched, or beaded, with a circlet of small sacs, and a fringe of long thin tentacles, which hang down waving and floating in the water like threads,—such is the general external appearance of the Medusa. The wall of this swimming-bell is so thin and filmy,

PLANT-ANIMALS, OR ZOOPHYTES. 161

that the internal arrangements can be readily seen. Within the body of the dome a tube hangs suspended, the lower end terminating in a mouth, the upper

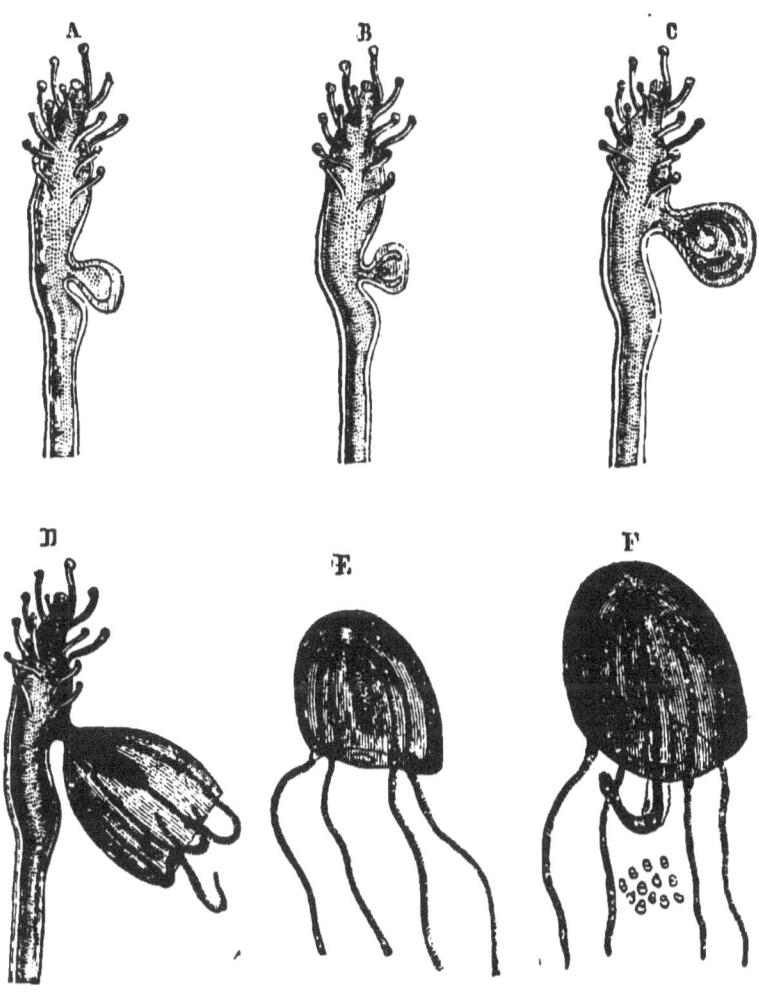

FIG. 26.—HYDROID POLYPE, PRODUCING A MEDUSA.

end united to a number of radiating tubes, which arch over, and pass downwards, to the edge of the swimming-bell, where they unite to a hollow circular ring, or canal. At the margin of the swimming-bell, and near the base of the tentacles, is one or more eye-spots, probably having the functions of an eye in higher organisms. The hollow cavity of the tube is the stomach, from which the digested food passes into the radiating canals, through which it circulates. The mouth is sometimes lobed, and sometimes fringed with a few short tentacles, to assist in the acquisition of food. This is a brief summary of the appearance, and structure, of the Medusa-like free zooid, in so far as relates to its movements, and method of living. All that concerns its other function, the perpetuation of the species, will be detailed further on.

"It would be difficult," writes Mr. Hincks, "to exaggerate in speaking of the beauty of these floating flower-buds, as they may well be called. The vivid tints which they often display, the gracefulness of their form, the exquisite delicacy of their tissues, and the vivacity of their movements, combine to render them singularly attractive. Frequently they are so perfectly translucent that their bubble-like forms only become visible in a strong light. In other cases the umbrella is delicately tinted, while the central tube (*manubrium*) displays the gayest colouring, and brilliant eye-spots glitter on the bulbous bases of the tentacles. To their other charms that of phosphorescence is often added; they are not only painted like the flower, but at night they are jewelled with vivid points of light, set round the margin of the bell, or one central lamp illumines the little crystal globe, and marks out its course through the water. Though individually minute, their numbers are so immense that they play an important part in the production of the luminosity of the ocean. The surface of the sea, for miles together, is often thickly covered with them; and on still, sunny days in autumn, certain species swarm in immense shoals off the coast. Like miniature balloons they float suspended in the water for awhile; then they suddenly start into motion, propelling themselves by a series of vigorous jerks, or casts, and at the same time contracting the tentacles into the smallest compass; then they become quiescent again, and sink slowly and gracefully, like parachutes, to the bottom of the vessel, some of the arms extended laterally, and the rest dependent. In all cases locomotion is effected by the pulsation of the swimming-bell."[1]

Increase by means of budding, which is so general in the "colony," is by no means absent from the free swimming polypes. It is not unusual when the true

---

[1] Hincks's "British Hydroid Zoophytes," p. xxx.

reproductive function is arrested, or in abeyance, for buds to be produced on different parts of the Medusa-like structure, sometimes on the central canal, but most commonly near the base of the tentacles, at the lower margin of the swimming-bell, and the buds thus produced originate free zooids, in all respects like the parent from which they sprung.

The reproductive elements are produced, in the free swimming polypes, either between the two membranes of the central pendulous tube, or they are enclosed in special cells, or sacs, seated on the radiating tubes. Sometimes they make their appearance before the liberation of the free zooid, and sometimes not until long after the zooid becomes free, the ovaries in one individual, and the spermaries in another, so that, typically, the free swimming polypes are either male or female.

It must not be supposed that all the genera produce the free polypes, for in some groups the sexual polypes always remain fixed, although this is regarded as being a less highly developed form of reproduction, and is subject to various gradations, some being simple, and others much more complex. In one sub-order the reproductive buds are borne on an off-shoot of the polypary, enclosed in an urn, or capsule, of similar substance to the residue of the horny skeleton. This is a large section, and contains some of the commonest, most interesting, and best-known species. The male and female are sometimes produced on the same polypary, but more commonly; perhaps, any given polypary is either male or female, as far as its reproductive buds are concerned, although Mr. Hincks thinks that the cases in which both male and female are mingled on the same shoots are much commoner than has been supposed.

After impregnation, the ova in the female capsules pass through intermediate stages, and at length become resolved into a granular mass, which by further development, becomes an elongated embryo (*planula*), which escapes from the capsule into the water, and temporarily enjoys a free existence, but it is the embryo which becomes free and *not* the sexual polype, as in the case of the veritable free swimming polypes. The liberated embryos have a double coating, or membrane, and the outer surface is almost always covered with vibratile cilia, or moveable hairs, by means of which their progress in the water is directed. Afterwards the body enlarges at one end, and a thin horny, or chitinous, film is formed over a portion of the surface, then its movements slacken, and at length cease, the cilia disappear, and the embryo becomes attached by the enlarged end, which flattens out into a disk, whilst the rest of the body stands erect in its centre. This is the first evident step towards the foundation of a new colony. The disk of attachment by-and-by becomes lobed and divided, forming the rudiments of the radiating base, and the whole structure is soon invested in a chitinous sheath or elementary polypary. In the course of time, as growth proceeds, the upper end of the attached embryo developes into a fixed polype, enclosed within a capsule, with a moveable lid, which is pushed off when the animal is mature. Then the initial step is complete, and the first individual, which for the time being constitutes the entire colony, is perfected; from this unit, by a series of successive buddings and branchings, the plant-like colony is gradually built up, in all respects resembling the parent, from whence the solitary embryo fell into the water, or was cast out into the world of waters, to make a home for itself, but a short time previously. And, in like manner, every polypary, with its hundreds of individuals, commenced, in such humble guise with a single polype.

Dr. Johnston has thus summarised what was known, in his time, of the growth of the polypary:—
" The ripe ovule, or bud, discharged from its matrix settles, and fixes itself to the site of its future existence, by minute fibres, which pullulate from the under side, while, from the opposite pole, a papillary cone shoots up to a height determined by the law which regulates the peculiar habit of the species. The upward growth is then arrested, and the apex becomes enlarged and bulbous. The structure of

this rudimentary shoot is at first apparently homogeneous, but very shortly the separation between the sheath and the interior pulp begins to be defined, and is made hourly more apparent by the pulp retreating inwards, becoming darker, and more concentrated. That portion of it in the bulbous top of the shoot goes on to further condensation and development; and as it enlarges, so in proportion does the horny cuticle that covers it expand apace, until it has gradually evolved into one or two cells, which are still closed on all sides. The dark body of the polype is apparent through the thin and transparent wall, and from its superior disk there are now to be seen some minute tubercles or knobs protruding, which, becoming, insensibly but steadily, more elongated, constitute the tentacles of the polype, now nearly ready for a more active life. By an extension of development, or by a process of absorption, not well understood, the top of the cell is at length opened, the polype displays its organs abroad, and begins the capture of its prey, for, unlike higher organisms, it is, at this the period of its birth, as large and as perfect as it ever is at any subsequent period, the walls of the cell having become indurated and unyielding, and setting a limit to any further increase in bulk. The growth being thus hindered in that direction, the pulp, incessantly increased by new supplies of nutriment from the polype, is constrained and forced into its original direction, so that the extremities of the tube, which have remained soft and pliant, are pushed onwards, the downward shoot becoming a root-like fibre, and the upper continuing the polypary, and swelling out as before, at stated intervals, into cells for the new development of other polypes."[1]

Undoubtedly the chitinous covering of the polypes, which serves as a protection to their delicate bodies, and a home in which to dwell, is gradually built up, as occasion requires, by absorption and elaboration, from the water of the ocean. These little builders construct their own houses, as the crab and the oyster construct their shells, from material held in solution by the sea-water. These structures may not be so imposing, or so substantial, as those of the architects who construct coral islands, but they are even more beautiful, and equally exhibit a facility for construction

—" Where every one,
By instinct taught, performed its little task ;
To build its dwelling and its sepulchre,
From its own essence exquisitely modelled ;
There breed, and die, and leave a progeny,
Still multiplied beyond the reach of numbers,
To frame new cells and tombs ; then breed and die
As all their ancestors had done,—and rest,
Hermetically sealed, each in its shrine,
A statue in this temple of oblivion ! "

---

[1] Johnston's " British Zoophytes," p. 9.

One most interesting feature in the history of these zoophytes is the phenomenon of phosphorescence which they exhibit, although they are not the only animals which contribute to the phosphorescence of the sea. One author writes:—" I lately had an opportunity of beholding this novel and interesting sight, of the phosphorescence of zoophytes, to great advantage when on board one of the Devonshire trawling-boats which frequent this coast. The trawl was raised at midnight, and great quantities of corallines were entangled in the meshes of the net-work, all shining like myriads of the brightest diamonds;"[1] and, subsequently, the same writer says:—"Numerous friends can bear witness to the exceeding brilliancy of the phosphorescent light emitted by a great variety of species, which I was in the habit of exhibiting to them. Once each week I received from the master of a trawling-vessel, on the Dublin coast, a large hamper of zoophytes in a recent state; in the evening these were taken into a darkened room, and the spectators assembled; I then used to gather up, with my hands, as much of the contents of the hamper as I could manage, and, tossing them about in all directions, thousands of little stars shone out brightly from the obscurity, exhibiting a spectacle, the beauty

---

[1] Hassall, in "Annals of Natural History," first series, vol. vii. p. 281.

of which to be appreciated must be seen, and one which it has been the lot of but few persons as yet to have looked upon. Entangled among the corallines were also numerous minute luminous Annelids, which added their tiny fires to the general exhibition."[1] It is unnecessary to add confirmation of what is now acknowledged as a fact, but the Rev. D. Landsborough may be quoted for an experience which befel him nearly fifty years since. "I brought from the shore in a pocket vasculum, or tin box, some zoophytes attached to sea-weeds, and laid the vasculum on the lobby table, till I should have leisure to examine them. When night came I put my hand into the vasculum to remove some of the zoophytes for inspection, and, on moving them, I found, to my surprise and delight, that they began to sparkle. Remembering what I had read, as I took them up I gave them a hearty shake, and they instantly became quite brilliant, like handfuls of little stars, or sparkling diamonds. To ascertain what were the zoophytes that emitted this phosphorescence, it was necessary to take them up singly by candle-light, and afterwards to make the experiment in the dark. One (*Obelia geniculata*) was very luminous, every cell for a few moments becoming a star ; and as each polype

---

[1] Hassall, in "Annals of Natural History," first series, vol. viii. p. 342.

had a will of its own, they lighted and extinguished their little lamps, not simultaneously, but with rapid irregularity, so that this running fire had a very lively appearance."[1]

It was in reference to this phenomenon that Crabbe wrote the following lines, so often quoted in connexion with this subject:—

> "While thus with pleasing wonder you inspect
> Treasures the vulgar in their scorn reject,
> See, as they float along, th' entangled weeds
> Slowly approach, upborne on bladdery heads ;
> Wait till they land, and you shall then behold
> The fiery sparks those tangled fronds infold,
> 'Myriads of living points ; th' unaided eye
> Can but the fire, and not the form descry."

The whole subject of the phosphorescence of the sea is of great interest, but hardly comes within our present scope. Very little of this general phosphorescence can be ascribed to our zoophytes,[2] although much, perhaps, to animals equally insignificant and humble. "I remember," writes an observer, "the admiration, not unmixed with wonder, (for then I knew not to what agencies the power by which water seemed suddenly to kindle and glow, as though turned to liquid fire, was to be attributed),

---

[1] "Annals of Natural History," December, 1841, p. 258.
[2] Readers are referred to a very interesting summary by M. de Quatrefages in *Popular Science Review*, vol. i. p. 275 (1862), entitled "The Phosphorescence of the Sea."

which I felt when first I viewed the beautiful phosphorescence phenomena of the ocean. Beautiful as this spectacle is, even in our own seas, in warmer latitudes, and in the Mediterranean, it is far more splendid; but, to be seen at all, it is necessary that the water should be disturbed in some way,—the slightest breeze curling the surface of the tranquil ocean, calls forth from its waters a flash of phosphorescent fire as it sweeps along—the wave, as it falls from the vessel's side, and breaks into ten thousand pieces, reveals innumerable globes of animated fire, suddenly called forth from the darkness which enveloped them—each stroke of the dripping oar scatters thousands of living gems around them, unequalled in brilliancy by the glittering of a kingly diadem—a golden path of light, increasing in breadth as the distance becomes greater, follows, like an attendant comet, the wake of the vessel urged onwards by the impelling wind—and the fisher's net, just raised to the water's edge, and laden with spoil collected from the secret beds, and hiding-places of the great deep, seems converted into a golden framework, set with precious jewels, by the presence of numerous zoophytes entangled in its meshes. Indeed, in whatever way the water is agitated the same beautiful appearance follows; if a little be placed even in the palm of the hand, and shaken, bright scintillations will be emitted; but, of course, the phenomenon will be more striking in proportion to the quantity of water put into commotion."[1]

The growth, or increase, of the polypary is comparatively rapid, for the size of the little animals that are concerned in building it up. Mr. Couch has expressed the opinion that a large specimen of one kind of *Sertularia* may be formed, under favourable circumstances, in the course of fourteen days.

Another observer mentions the fact of the bottom of a boat which became covered in a fortnight. Some of the species undoubtedly continue through several seasons, and increase in size from year to year, whilst others have a much briefer career, and some only last during a single season. Those kinds which attach themselves to rocks are considered to be more persistent than those which affect sea-weeds. In perennial species the polypes are much reduced in number during the winter, reappearing again in the spring. Although several kinds have been known to thrive with apparent vigour in aquaria, yet the rate of increase, under such artificial conditions, would hardly be equal to what it would be in a state of nature. The enormous rate at which many of them multiply, and establish new colonies, might almost be anticipated from the large number of individuals

---

[1] "Annals of Natural History," first series, vol. viii. p. 343 (1842).

that are included in a single thriving colony. An American naturalist says, that he has seen one species discharging embryos as early as March, and as late as the middle of September, "during all which time thousands were continually shed, and consequently thousands of new colonies established, their multiplication being so great, during a favourable season, that the rocks literally appeared clothed with the yellow stems, and rose-coloured blossom-like bodies, of these flower-animals."

It being the object of this chapter to indicate generally, the structure and habits of the little builders who construct the horny homes in which they dwell, as "Toilers in the Sea," it is unnecessary to particularise the different species, or point out their characters, whereby one may be known from the other, since those who are desirous of continuing the study further will not expect to find such information in one short chapter, and, moreover, there is already an excellent text-book for that purpose, by the Rev. Thomas Hincks, to which we have already alluded.

"Amongst the rejectamenta which strew the shore after a fresh breeze and a rough tide, there are few things more likely to arrest the eye, and win the admiration, even of the least observant, than the zoophytes, which mingle their light and flexile forms with the tangled masses of weed, or lie in graceful tufts upon the sand. With the uninitiated, their plant-like configuration passes as conclusive evidence of their vegetable nature ; and the non-scientific world now, like the men of science of a century ago, receives the doctrine of their animality with incredulity. And, indeed, we cannot wonder at the scepticism ; for so complete is the imitation of vegetable forms in these beings, and so vegetative in many respects is the fashion of their life, that there is nothing to suggest a doubt, on a slight and superficial acquaintance, as to their affinities. The skeletons only of the zoophyte commonly fall in the way of the amateur collector on the shore, and these, intermingled with the algæ, and resembling nothing so much as miniature shrubs, and trees, offer no sufficient clue to the interpretation of its history."[1] It may suffice to say that the best known of these zoophytes are those which belong to the sub-order *Thecaphora*, in which the polypes and reproductive buds, are enclosed in protective cases, and especially the Sertularians, which belong to the families named *Sertulariidæ* and *Plumularidæ* (fig. 27). These names will convey no ideas to those unacquainted with the small details which make up the systematic characters of these little animals, but they

---

[1] "The Sertularian Zoophytes of our Shores," by the Rev. T. Hincks, in *Popular Science Review*, vol. viii. p. 223 (1869) ; see also " The Hydroid Medusæ," in *Popular Science Review*, vol. xi. p. 337 (1872).

FIG. 27.—HYDRALLMANNIA FALCATA.

may indicate the direction in which further information is to be sought (figs. 28 to 32).

Amongst all the minute organisms which make their homes in the "deep, deep sea," there are none more beautiful, or instructive, especially when seen under a low power of the microscope, than the "plant-animals" or zoophytes. Other colonies may have a history almost as strange, but they do not offer the same facilities for observation as do these dwellers in almost transparent houses, through which much of their domestic economy may be studied. They are always favourites with the sea-shore collector, but the empty cases are but shadows of the living animals, as seen in their native element. "There must always" —and here again we quote Mr. Hincks—"be a certain fascination in a history which tells us of animals composed of multitudes of individuals (zooids) living an associated life, and so combining as to produce the most graceful plant-like structures—vegetating like a tree—putting forth thousands of polypites, like leaves, each a provider for the commonwealth—putting forth also a company of buds, charged with the perpetuation of the species, ripening in transparent urns, and scattering their winged seeds broadcast, or sent forth, moulded and painted by the highest art, like fairy emigrant ships freighted with young life, to colonise distant seas."[1]

---

[1] "Zoophytes: the History of their Development," in *Quarterly Journal of Science*, vol. ii. No. 7, p. 415.

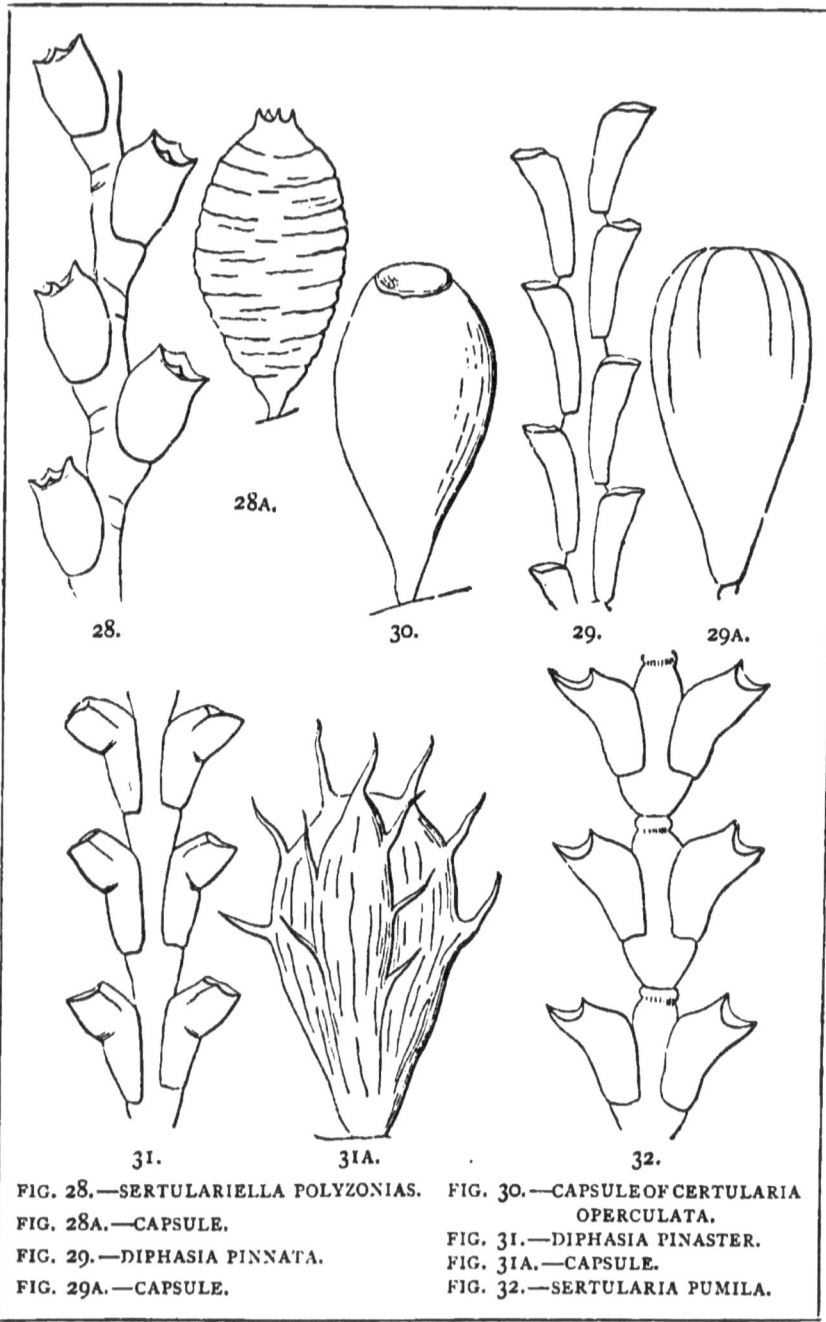

FIG. 28.—SERTULARIELLA POLYZONIAS.
FIG. 28A.—CAPSULE.
FIG. 29.—DIPHASIA PINNATA.
FIG. 29A.—CAPSULE.
FIG. 30.—CAPSULE OF CERTULARIA OPERCULATA.
FIG. 31.—DIPHASIA PINASTER.
FIG. 31A.—CAPSULE.
FIG. 32.—SERTULARIA PUMILA.

We cannot close this history better than in the words of Ellis, one of the oldest of our writers on the Corallines:—" And now, should it be asked, granting all this to be true, to what end has so much labour been bestowed in the demonstration? I can only answer, that as to me these disquisitions have opened new scenes of wonder, and astonishment, in contemplating how variously, how extensively, life is distributed through the universe of things; so it is possible, that the facts here related, and these instances of nature, animated in a part hitherto unsuspected, may excite the like pleasing ideas in others; and in minds more capacious and penetrating, lead to farther discoveries, farther proofs (should such be wanting), that One infinitely wise, good, all-powerful Being has made, and still upholds, the whole of what is good and perfect; and hence we may learn that, if creatures of so low an order in the great scale of nature, are endued with faculties that enable them to fill up their sphere of action with such propriety; we likewise, who are advanced so many gradations above them, owe to ourselves, and to Him who made us, and all things, a constant application to acquire that degree of rectitude, and perfection, to which we also are endued with faculties of attaining."

CHAPTER VI.

SEA-FAN MAKERS.

AMONGST the most common objects of the seashore, all around our coasts, are those strange looking, scarcely handsome, lumps of dingy, cold, soft substance, known as "dead man's fingers," and that is all which most people know, or care to know, about them; and yet we must make their better acquaintance, because the structure of the living animal will serve as a key to that of the far more ornamental, but less common objects, called "sea fans." The polyp-home, or polypidom, resembles a compact sponge; sometimes it is only a thin crust of less than a quarter of an inch, but usually it rises in conoid, or finger-like masses, of various sizes, and are either simple or lobed. When it is simple the fishermen of the coast call it "cow's paps" (fig. 33), but when divided into lobes, like fingers, it bears the name of "dead man's toes," or "dead man's hands." The outer skin is tough and leathery, studded all over with star-like figures, which, if attentively examined, are seen to be divided into eight rays, indicating the number of the tentacles of the polyps, which issue here. The body of the polype is, as it were, enclosed in a transparent vesicular membrane, dotted with many minute calcareous grains, and marked with eight white longitudinal lines or septa, which, stretching between the membrane and the central stomach, divide the intermediate space into an equal number of compartments. These lines not only extend to the base of the tentacles, but run across the disk, and terminate in the central mouth. The tentacles are short, obtuse, fringed on the margins, and strengthened at their base by numerous linear straight crystalline spicules. From the base of the white longitudinal lines an equal number of white tortuous glandular filaments depend, hanging loose in an abdominal cavity placed underneath the fleshy cylindrical stomach, and continuous with the canals. The polyp-cells are oval, placed just beneath the skin, and are the terminations of the long water-bearing canals which run through the whole polype-mass. These canals divide, as they proceed, into branches, which diverge towards the circumference, and there expand into cells; they have strong, per-

FIG. 33.—"COW'S PAP" (*Alcyonium digitatum*).

haps muscular, coats, and are filled with a much less consistent matter than that of the body of the polyp itself. Hence it appears that many polypes communicate together, and form a compound animal. The space between the tubes is occupied by a loose fibrous tissue, with the interstices filled with a transparent gelatine, in which lie numerous irregular

FIG. 34.

spicules. These spicules are mostly of the form of a cross, and toothed on the sides, formed of carbonate of lime (fig. 34), and have no organic connexion either with the tissue or the tubes. The ova are contained in the polyp tubes; they are white at first, but at length become of a scarlet colour, opaque, globose, and about the size of a grain of sand; ultimately they are expelled from the mouth.[1] The polyp-masses are of a greyish white

---

[1] "History of British Zoophytes." By George Johnston, M.D. (London, 1847.) Vol. i. p. 175.

colour, or almost that of wet sand, or of a more or less bright orange. We have often seen them adhering to the scallop-shells exposed for sale in fishmongers' shops, but, of course, in a dead condition. When dredged up, fresh from the sea, and placed in a vessel of sea-water the polyps expand, and then the mass, which before was so uninteresting, becomes quite enchanting.

It will be remembered that the Actinoid polypes, of which the sea-anemones are the type, have their tentacles and internal septa a multiple of *six*, but the Alcyonoid polypes, or those of which the "sea-paps," or "dead man's fingers," is the type, have *eight fringed* tentacles. To this group belong also the sea-pens (*Pennatula*), the sea-fans (*Gorgonia*), the organ-pipe coral, and the well-known "red coral." The animals in all these are very similar, and have the common distinguishing mark of eight fringed tentacles, but the sea-paps are flexible, containing spicules of carbonate of lime scattered amongst their tissues, and without the hard rigid central axis which is present in the sea-fans, except in one family, to which the pipe corals belong, in which the animals secrete slender tubes of carbonate of lime. In the sea-fans, on the contrary, there is a hard, horny axis, of greater or less thickness. In some cases the axis is so slender that the coral (if we may so call it) does not stand erect, but hangs pendulous from the rocks.

The great sea-fan of the West Indies often grows to a yard in height and breadth, the branches and twigs meet and coalesce, so as to form a regular net-work, but in many other species the branches coalesce much less frequently, or not at all, and then they resemble clusters of slender twigs, or miniature shrubs or trees. The general colour is often very bright, either red or yellow, or violet, and must have an attractive appearance, if they could be seen growing in their ocean depths, attached to rocks, or to the sides of coral reefs. The outer surface of the stems and branches of the sea-fans is like a bark, and hence often called the *cortex*, which may be peeled off like the bark of a twig, leaving the hard and firm axis intact. This bark is a continuous layer of the united polypes mixed with minute granules, or spicules, of carbonate of lime (to be described more fully shortly). The outer surface is commonly smooth, but is sometimes covered with small prominences ; where the latter are present they are surmounted by the orifice, or oblong puncture, through which the animal protrudes itself during its life (fig. 35).

Devotees to the use of the microscope, especially those who employ it chiefly as a source of amusement, usually include amongst their "beautiful objects" preserved and prepared specimens of "spicules of Gorgonia," which, being interpreted, implies that they are the little granules of carbonate of lime secreted by the animals in the cortex, or bark, of the sea-fans. They are called "spicules" because they are not amorphous or crude granules, but beautifully symmetrical shapes of a certain definite type, sometimes

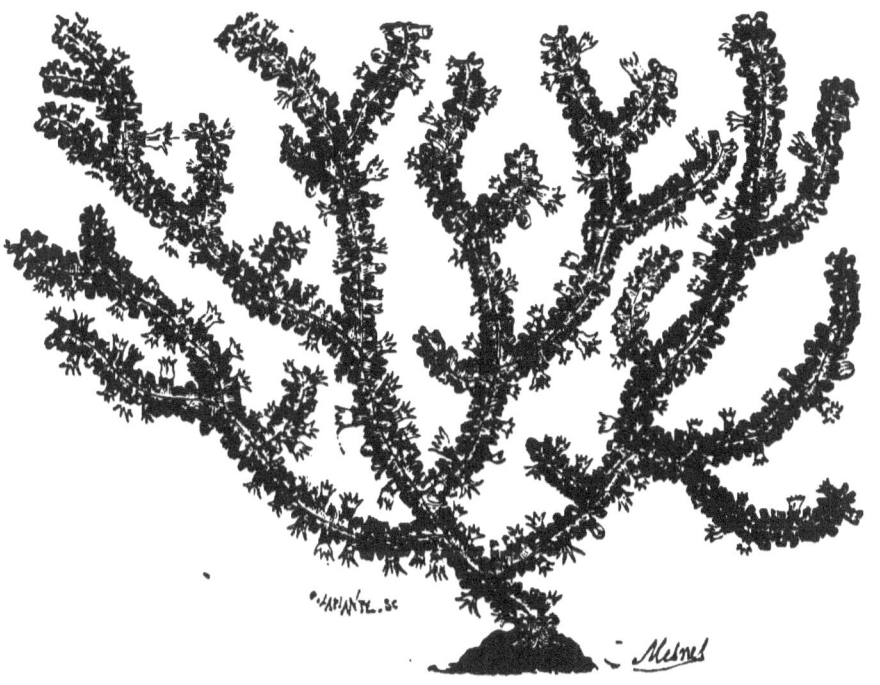

FIG. 35.—GORGONIA, OR SEA-FAN.

white, but usually yellow, scarlet, crimson, violet, purple, or, in fact, of the same colour as the sea-fan itself, for it is to these coloured spicules that the characteristic colour of the Gorgonia is due. Sometimes the spicules are shaped like a cross, sometimes like a club, sometimes oval or oblong, with the surface ornamented with more or less complex bosses or projections (Plate IV.). Certain authors believe that they can determine the genus, even if not the species of the family of Gorgonia, from the form of their calcareous spicules, which, at the least, must be accepted as implying that, although the forms vary in the spicules of different species, they are at the same time constant in each of the species.

It may be intimated here that the spicules of these organisms may be easily removed from the bark, and cleared of all extraneous matter, so as to be mounted permanently as "objects for the microscope." A small portion of a branch, or of the cortex from a branch, may be boiled in a test-tube, with *liquor potassæ*, for a minute or two, and when cool and settled, the sediment may be washed by shaking in water, allowed to settle, the water poured away, and more water added, shaken again and again allowed to settle, as often as need be, and the final sediment will be found to be the clean spicules which may be mounted dry, or immersed in Canada balsam. Some prefer to mix spicules of different forms and colours together, others to keep each kind separate and distinct; let every one be persuaded in his own mind.

M. Valenciennes, who wrote a monograph of this family, has classified the forms of spicules in the following manner:—

(1). The spicules (he calls them *sclerites*, to distinguish them from the spicules of sponges) have two small circles of tubercles at a distance from one another on a short axis; the tuberculate extremities resemble a small branch of cauliflower. This kind is found in the common red coral.

(2). These spicules are spindle-shaped, with four or six circlets of tubercles.

(3). In the clubbed spicules a single extremity is dilated, and furnished with ridges like some ancient maces.

(4). The muricated spicules have four or several points, and entirely covered with spines.

(5). The last form consists of larger or smaller scales, more or less covered with small spines.[1]

He comes to the conclusion that although useful in the characters of the different species, yet that these spicules cannot be employed for the distinguishing of one genus from another, because more than one kind of spicule occurs in the same genus.

Milne-Edwards, and after him, Kolliker, succeeded to Valenciennes, and brought nearer to perfection what he had commenced, especially in the matter of classification. The most noteworthy direction in which Kolliker diverged from Valenciennes, appears to be the

---

[1] "Abstract of Monograph of Gorgonidæ," by M. Valenciennes, in "Annals of Natural History," vol. xvi. (1855) p. 177.

acceptance of the form of the spicules as of higher importance, even in the determination of genera. In this view, also, Mr. W. Saville Kent concurred to the extent even of going further than Kolliker had done, for he says :—" The facts which have been eliminated will, I think, suffice to demonstrate beyond doubt what a highly important element the calcareous spicules represent in our appreciation of the generic characters of the *Gorgonidæ* (sea-fans). The study of the group from this point of view must, however, be considered quite in its infancy ; but the time has arrived when zoologists must no longer be content with the characters afforded by general contour, or by the examination of the dried polyparies only."[1]

The five types of spicules enumerated by M. Valenciennes have not been deemed sufficient by succeeding observers, who have come to the conclusion that the material at his disposal was insufficient, and that more definite and typical forms of spicule have since been met with, with which he was unacquainted, hence that the number of leading forms must be increased, and hereafter probably will be increased, to an extent almost equal to the number of genera based upon them.

---

[1] "On the Calcareous Spiculæ of the Gorgonaceæ," by W. S. Kent, in *Transactions Royal Microscopical Society*, Feb., 1870.

It has been intimated, that in the "sea-fans" there is to be found a horny axis, covered with a fleshy bark, or cortex, from which the polypes protrude. This axis, in the majority of cases, is accurately characterised as horny, for it is composed of a substance very similar, but not identical with that which constitutes the horns and hoofs of animals. Still, there remains a small group in which the axis is not absolutely horny, but partly of the horny substance and partly of carbonate of lime. Of course, both of these substances are secreted and deposited by the polyps.

A "sea-fan," or Gorgonia, therefore, is a body which is formed by the union of numerous polypes upon a common body, enveloped by a gritty, fleshy substance, containing spicules of lime, and surrounding another tree-like secretion, which is its axis, formed of a substance resembling horn, or horn mixed with carbonate of lime, the polypes themselves being furnished with eight arms, or tentacles, the margins of which are fringed. Bearing in mind these facts, it will be seen readily what the "sea-fans" have in common, and wherein they differ from, their nearest associates.

It is difficult to enter with precision into the details of the structure of these minute animals, and their secretions, without being exposed to the charge of being "heavy" and "dry," but some excuse, even for such a failing, may be found in the fact that so many persons have in these latter days habituated themselves, not only to cultivate the use of the microscope, but also to carry it with them, in their holiday excursions to the sea-side, and that such persons desire, above all things, abstracted and compact information on marine objects (as well as many other subjects) which shall introduce them fairly, not only to the common, but some of the uncommon objects of the seaside, or the nearest representatives that they can procure of the inhabitants of the tropical seas. Those who cannot collect the living "sea-fans" of warmer climes will not be disgusted, or discontented, with their humble cousins, the "dead man's fingers" of our own shores.

The Tubiporine, or Organ-pipe coral, is a favourite object in all museums and collections of marine productions, and is of interest as belonging to this section, although of such diverse external appearance, consisting of a great number of hard purple-red tubes, about the thickness of a straw, standing nearly parallel to each other, like miniature organ-pipes, and united at short distances by thin plates of the same coral-like substance (fig. 36). The horny axis of the sea-fans is absent, but is compensated for by the rigid tubes, which appear to be formed gradually at the upper end, by the deposition and consolidation of the calcareous spicules secreted by the animals.

Professor Perceval Wright found masses, as much as 2 ft. in diameter, in the Seychelles, although, more commonly, in irregular lumps of about 12 in. in circumference, and from 2 in. to 4 in. in height. We are indebted to him for the description of the animals which inhabit, and construct, these tubes, as he had the good fortune of seeing them in the living state. " The polyp," he says, "consists of eight pinnate tentacles, each tentacle with from fifteen to seventeen pennæ on each side; these tentacles are thickly studded with spicules of an oval shape, flat, somewhat longer than broad; and are met with all over the tentacle, down the centre of which there is one compact row, forming, as it were, a mid-rib; they are often slightly compressed in the centre, so as to form a figure of eight. In the centre of the tentacles is the mouth, with a slightly raised circular lip. When the polyp is alarmed the tentacles are first closed together, and then the polyp sinks down quite into the tube; as it becomes more completely retracted it draws in after it the uppermost portion of the tube itself, inverting this, and folding it in until the open mouth of the tube is thereby completely filled. It

FIG. 34.—ORGAN CORAL
(*Tubipora musica*).

is, of course, only the yet spicular and not the solid portion of the tube that is thus inverted; and the folds thus formed equal in number the tentacles. I have more than once traced these spicular portions up to the very base of the tentacles, where the fusiform spicules end, and the characteristic tentacle and body spicules commence; these spicules thus forming a series of triangular spaces, the bases of which join on with the hardened edge of the tube, and the apices are situated at the base of each tentacle. The spicules secreted by this portion of the ectodermic layer are of several sorts: first, the warty fusiform spicule, so commonly met with in the Alcyonidæ; these spicules will be found in all stages of growth, and of coalescence; thus, at the upper portion of the edge of the tube, where it is non-retractile, the calcareous tissue will be found to consist of a series of them, partially joined together, and making a kind of coarse open network, which, on being macerated in caustic potash, does not fall to pieces; but the retractile portion, on being subjected to the same treatment, breaks up into a mass of minute individual spicules. The red colouring matter would appear not to reside in these latter spicules, for those that I have examined are colourless. A second form of spicule is met with in the retractile portions of the tube, which, I think, might be called 'shuttlecock.' While all the forms of spicules (fig. 37) met with seem to occupy certain definite portions of the ectodermic layer, yet there is an evident gradation between them, from the smooth,

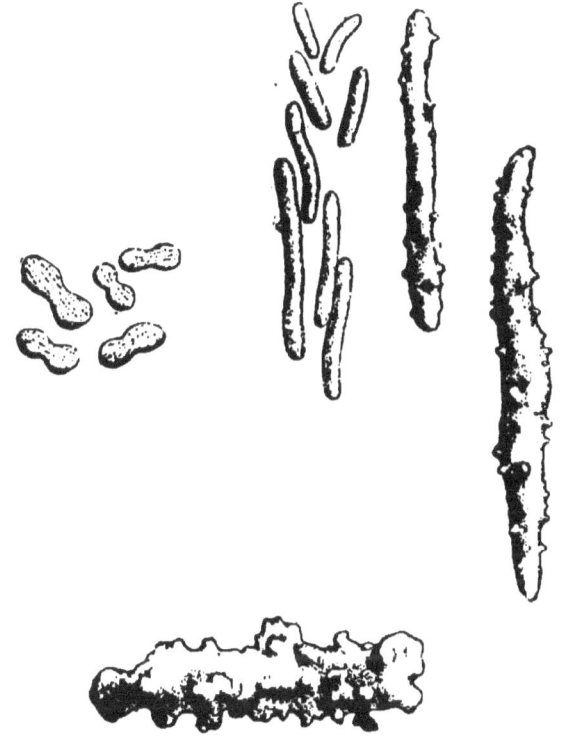

FIG. 37.—SPICULES OF ORGAN CORAL.

fusiform spicule to the most irregularly warty forms, which leads naturally to the inference that all these forms are but different stages of growth, by the aggregation of new calcareous material, until the solid tubular structure, so long known to us, is at length reached.

"The mouth, which is circular, is distinctly marked, and leads into the stomachal cavity, which is small; the stomachal cavity is separated by a thin and delicate membrane from the general body cavity. I have not been able to determine with exactness the number of openings between these two portions. The ovaries are in the general cavity, and are invested by a delicate membrane, which is continued in the form of eight mesenteric bands to the body of the tube.

"The structure of the skeleton is certainly not crystalline, as some have suggested. Fusiform spicules are secreted by the ectodermic layer; these spicules around the base of the tentacles are of a white colour, and in many cases are simply fusiform, not warty; but those at a little distance from the base of the tentacles not only assume a light red colour, but become crowded over with warty excrescences, and there is then to be found a gradual growing together, and consolidation, of those around the edge of the tube, that is, where this is formed. In the case of a young bud, there is at first no tube, the spicules having not yet become coalesced; they are here simply placed side by side.

"The polyp certainly can, and does constantly, add to the height of its tube; or, in other words, the spicules are being constantly consolidated into the tube, and the tube thus increases in height. In some cases I have been able to trace the mesenteric bands, which attach the lower portion of the body of the polyp to the walls of the skeleton tube, as far as the second external septum in depth; and it is very evident that, as the outer walls of the tube become consolidated, not only does the tube become elongated, but the polyp elevates itself at the same time in the tube."[1]

As the "Organ-pipe coral" is an inhabitant of warm latitudes, and has seldom been observed or studied, even there, in the living state, the foregoing particulars of its animals will have a special interest. This is one of the objects with which so many are familiar, in its dead condition, or when nothing remains save the skeleton, but concerning the inhabitants of which so little was previously known that even now it may be assumed that information is far from complete. Indeed, its reproduction and development, which would entail observation over a considerable period of time, is still guessed at rather than demonstrated. From analogy it may be assumed that in these particulars it does not differ greatly from its imme-

---

[1] "On the Animal of the Organ-pipe Coral," by Prof. E. Perceval Wright, in "Annals of Natural History," vol. iii. (1869) p. 377.

diate allies, and that those authors are not far wrong who claim for the *Tubipora* a near kinship to the ordinary "red coral of commerce."

Until about 1727 the red coral was held to be a marine plant, but now it is acknowledged as an animal production, closely related, in its scientific affinities, to the sea fans.

It has been shown by a recent author that the beautiful red coral, with which most of us have been acquainted from infancy, bears in its Latin name (*Corallium rubrum*) the poetical designation of "the red daughter of the sea."[1]

It is found in the Mediterranean, and the Red Sea at various depths from ten to nearly a thousand feet. "Each stalk of coral" writes Moquin - Tandon, "resembles a pretty red leafless shrub, bearing little delicate star-like flowers. The stalks of this little tree are common to the association, and the flowers are the polypes. These arborescent formations generally hang from some shelf, and so grow downwards, and not like ordinary vegetation. They are found growing together in bushes, or copses, or spreading out, as we have said, into veritable forests. The stems have a soft reticulated cortex or bark, which is full of little cavities, permeated by a milky

---

[1] *Korallion* from *Korē*, a daughter, and *alos*, the sea; Latinised *curalium*, or *corallium*, with *rubrum*, red.

fluid; these are the chambers of the members of the association. The cortex is full of little hard bodies, called spicules (fig. 38). Beneath the crust is the coral, properly so called; it is as hard as marble, its surface is remarkable for its stripes, its colour is a beautiful red, and it is so hard that it is capable of receiving a high polish. The

FIG. 38.—SPICULES OF RED CORAL.

polypes are composed of a tubular body, enclosed in a cortical cell. That part which appears beyond the walls of this cell is cylindrical, and is crowned by eight little arms, which spread themselves out like the petals of a flower. These arms are flat and wide at the base, but gradually tapering to a point. The edges are furnished with short hollow barbs. When expanded they are exactly like a beautiful white and semi-transparent flower, having eight petals fixed upon a mammal bud; when closed they have the appearance of an urn."[1] It will be observed that the red coral differs considerably in its animals from the corals of the coral reefs. In the latter case the animals are either of the Anemone type, or what are termed in scientific language Actinoid polyps, or, as in the Millepores, of the hydroid type; whereas in the red coral they are of the type which prevails in "dead man's fingers," and the "sea-fans," that is of the Alcyonoid type, hence "precious coral" is more nearly related, in its structure, to the Gorgonias than to the rock corals. Probably the general reader will not at once appreciate the differences, or their importance, so readily as the zoologist, although we have placed them in different chapters in this volume, in order to emphasise that difference and prevent the confusion of the two kinds of substance to which the name of "coral" has been given.

The structure, and reproduction, of the red-coral polyp has been studied more exhaustively by M. Lacaze-Duthiers than by any one else, before or since, and a summary of the facts which he has made known will be of interest, to compare with those we have given elsewhere of the rock corals.

---

[1] Moquin-Tandon: "The World of the Sea," London, 1869, p. 109.

From him we learn that the members of a coral association are either males, females, or hermaphrodites. Ordinarily, the greater number of polypes on one branch are of the same sex; one branch containing almost exclusively males, another females; the hermaphrodites are the least numerous. The coral polyp is viviparous—that is, the eggs become embryos before they leave the parent. The eggs have long slender pedicles; they come out of the thin layers which line the digestive bag; they are spherical, opaque, and of a milk-white colour. They detach themselves by breaking their pedicles, and thus fall into the central cavity of the polyp, a cavity which serves as a stomach and an incubating pouch. Here two very different processes are in action,—the one dissolving the food and nourishing the animal, the other developing and producing a new creature. In due time the eggs lengthen, and vibratile cilia appear. As soon as they are born—that is, vomited—a pore opens at one of the extremities, which is destined to become the mouth. Then they assume the form of a whitish semi-transparent worm. These larvæ swim in all directions, with the greatest agility; they rise and sink in the vases which contain them, always swimming with their thicker extremity in advance, carrying their mouths in the rear, so that they butt against anything which happens to be in their way. They have a tendency to become fixed, like their parents, and the mode of their progression greatly favours this result; thus the peculiarity of their motion is liable to shorten the period of their liberty, in facilitating their adherence to any object with which they come into contact, by that part of their body which afterwards becomes the base of the polyp. When it adheres the polyp has reached another stage of its existence. As soon as it becomes fixed it changes its worm-like appearance, and thickens, gaining in breadth what it loses in length, thus shortening, and becoming discoid. The thin extremity, which carries the mouth, gradually folds itself back, by successive stages, until the larva assumes the shape of a pin-cushion, at the summit of which is the mouth. Around this orifice the rudiments of the eight tentacles begin to appear, which soon cover it with a pendant festoon. The fixed larva thus becomes the founder of a large colony. Buds form on the axis, and develop themselves into a whole nation of corals. The young polypes, just hatched from the egg, are, as we have seen, utterly different from their parents. They must undergo many metamorphoses before they reach their perfect state. These metamorphoses are just the reverse of those through which insects pass. The chrysalis lies immovable, but finally becomes a butterfly. With the coral, the larva has the power of locomotion, whereas the full-developed polyp is fixed. There is not, perhaps, in nature a single law whose reverse is not also found in operation.

We find in these animals, just as we do in vegetables, stalks and branches covered with real bark; but the stems, or axes, of the animals are horny or calcareous, while those of the vegetables are herbaceous and woody. In each case the tissue is more or less solid, striated, or fluted, and formed of concentric layers; and, moreover, the cortex of the coral is spongy, and somewhat soft, like the bark of a vegetable production. The knobs represent the buds; the polypes, the flowers; the feelers open themselves out in rosettes as petals, forming an animated corolla, which opens and shuts alternately (fig. 39). In the polypier, the individuals which contribute to the growth of the whole are situated at the extremities of the axis, or upon its sides, a position similar to that of the leaves and flowers in a plant.

FIG. 39.
A RED CORAL POLYPE.

And the final resemblance is found in their modes of reproducing their species. The corals, and the vegetables, are each produced by isolated and individual grains, eggs, and seeds, which separate themselves from a bunch, and proceed to develop themselves, producing a colony, of which the members remain grouped.[1]

A far more extended and explicit account is given of the results of M. Lacaze-Duthiers' researches, by the Editor, in *Popular Science Review* (1865), but the foregoing is sufficient for our present purpose. The article in question may be consulted by those who desire to investigate the subject further. A section of the coral axis, or corallum, the red coral of commerce, may be cut and reduced to thinness, sufficient for examination by the microscope, when it will be perceived that "the central portion of the stem is occupied by a homogeneous coloured mass, of mineral character, round which there appears to be a ribbon-like fold of more highly tinted matter, which does not form a circle, but some irregularly oval figure; between this and the circumference one sees alternate concentric rings of strongly and faintly tinted material, which are divided by numerous radiating dark lines of extreme tenuity. These various appearances may be thus accounted for: the central, irregularly folded, ribbon-like mass is the first hard part developed by the young egg-polyp; this had the mineral matter deposited all

---

[1] M. Lacaze-Duthiers, "Histoire Naturelle du Corail," Paris, 1864; and Moquin-Tandon, "Le Monde de la Mer."

round it year by year, but the deposit went on with greatest rapidity in the greatest depressions, and hence, after a while, a circular outline was produced ; the several rings indicate the successive depositions, and the dark radiating lines point to the grooves, in which in earlier days the central large canals surrounded the stem."[1]

It has been estimated that the annual value of the coral dredged in the Mediterranean amounts to nearly £180,000, but M. Lacaze-Duthiers thinks that this is nearly three times as much as it really should be, and that in 1860 probably the total value did not exceed £60,372. At a rough estimate, average coral realises about twenty shillings per pound to the merchants, some kinds being worth considerably less, and the choicest kinds very much more. A delicate pink variety, being the scarcest, obtains the highest price, and is greatly esteemed by the Italians. It is the opinion of many that the coral fishery is in a state of gradual decay, and that the coral-beds are yearly becoming more and more exhausted, so that, unless something is done, either by artificial culture or otherwise, this industry will, at no very remote period, be numbered with the things of the past.

---

[1] Dr. H. Lawson, "Natural History of Red Coral," in *Popular Science Review* (1865), p. 72.

Some years ago, a writer in the *Athenæum* gave a graphic account of the coral fishery, as pursued in the Mediterranean, from which the following paragraph may be read with interest. It contains chiefly the mode in which this fishery, as an industry, is conducted, or at least as it was pursued when this account was written.

Torre is the principal port in the south of Italy for the vessels engaged in the coral fishery, —about two hundred vessels setting out from hence every year, about June. They have generally a tonnage of from seven to fourteen tons, and carry from eight to twelve hands; so that about two thousand men are engaged in this trade, and, in case of an emergency, would form a famous *corps de réserve*. They generally consist of the young, and hardy, and adventurous, or else the wretchedly poor, for it is only the bold spirit of youth, or the extreme misery of the married man, which would send them forth upon this service. For two or three months previously to the commencement of the season, many a wretched mariner leaves his starving family, and, as a last resource, sells himself to the proprietor of one or another of these barks, receiving a caparra (earnest money), with which he returns to his home. This, perhaps, is soon dissipated, and he again returns, and receives an addition to his caparra; so that, when the time of final departure arrives, it not unfrequently happens that the whole of his scanty pay has been consumed, and the improvident or unhappy rogue has some months of hard labour in prospect, without the hope of another grano of compensation. Nor does the proprietor run any risk in making this prepayment; for as the mariner can make no engagement without presenting his passport, perfectly *en règle*, he is under the surveillance of a vigilant police. The agreement between the parties is made from the month of March to the Feast of San Michaele (29th September) for vessels destined for the Barbary coast, and from March to the Feast of the Madonna del Rosario (October 2nd) for those whose destination is nearer home. Each man receives from twenty to forty ducats, according to his age, or skill, for the whole voyage; whilst the captain receives from one hundred and fifty to four hundred ducats, reckoning six ducats to one pound sterling. These preliminaries being settled, let us imagine them now on full wing,—some for the coast of Barbary, and others for that of Sardinia, or Leghorn, or Civita Vecchia, or the islands of Capri, San Pietro, or Ventotene, near which I have often seen them, hour after hóur, and day after day, dragging for the treasures of the vasty deep. On arriving at the port nearest to the spot where they mean to fish, the "carte" are sent in to the consul, which they are compelled to take again on return. A piastre is paid by each vessel for the magic endorsement of his Excellenza, another to the druggist, and another to the medical man; whilst the captain, to strengthen his power, and to secure indemnity, in case of some of those gentle excesses which bilious captains are sometimes apt to commit, has generally on board some private "regalo" for his consul. The next morning, perhaps, they push out to sea, and commence operations, — not to return that evening, or the next, or the next, but to remain at sea for a fortnight, or a month at a time, working night and day without intermission. The more humane captains allow half their crews to repose from Ave Maria to midnight, and the other half from midnight to the break of day; others allow only two hours' repose at a time; whilst some, again, allow no regular time; "so that," said a poor mariner to me, "we sleep as we can, either standing, or as we haul in the nets." Nor do they fare better than they sleep; for the whole time they have nothing — literally nothing — but biscuit and water, whilst the captain, as a privileged person, has his dish of dried beans or haricots boiled. Should they, however, have a run of good luck, and put into port once in fifteen days or so, they are indulged with a feast of maccaroni. These privations make it rather rough work, it must be confessed, for a mariner, especially when it is remembered that it lasts seven months; but if to this be added the brutality of the captains, whose tyranny and cruelty, as I have heard, exceed anything that has ever been recounted to me before, we have a combination of sufferings which go far to justify the description given to me of this service by one engaged in it, as being an 'Inferno terrestre.'"[1]

Having had occasion to mention the "sea-pens" in the early portion of this chapter, and dismiss them abruptly, it may be permitted to return to them for a page or two, especially as one species is common enough in our Northern seas, although rare in the South, and it may be the good fortune of some of our readers to make their personal acquaintance. These colonies are not fixed, as are the sea-fans, but, on the contrary, are free swimmers. In shape they bear some resemblance to a feather or pen, with a common axis, and lateral barbs or pinnules, which bear the polyps; these pinnules are longest near the centre, and shorter towards either end (fig. 40). The polyps are arranged in transverse rows along the outer

---

[1] The "Technologist," vol. ii. p. 22 (1862).

and inner edge ; they possess eight tentacles, and these are ciliated on one side. According to recent researches, the polypes of each pen are either all male or all female, so that all the members of one community are of the same sex (fig. 41). The British species is called by the fishermen the "cock's-comb," but this is scarcely so expressive as sea-pen. "It is from two to four inches in length, and of a uniform purplish-red colour, except at the tip or base of the stalk, where it is pale orange-yellow. The skin is thickish, very tough, and of curious structure, being composed of minute crystalline cylinders, densely arranged in straight lines, and held together by a firm gelatinous matter or membrane. These cylinders (spicules) are about six times their diameter in length, straight and even, or sometimes curved and bulged, and of a red colour, for the colour of the zoophyte is derived from them, and they are accordingly less numerous where the purple is faint or defective. The stalk is hollow in the centre, and contains a long slender bone which is white, smooth, square, and tapered at each extremity to a fine point. It does

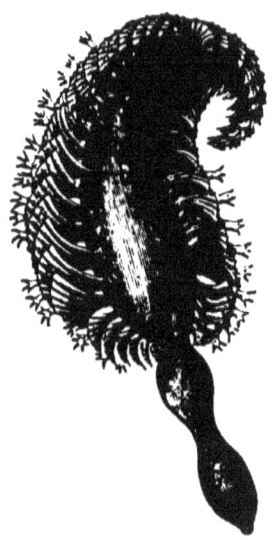

FIG. 40.
THE THORNY SEA-PEN.

not reach the whole length of the stalk, but before it reaches either end the point is bound down and bent backwards like a shepherd's crook."

It has long been known that the sea-pens are phosphorescent, and, indeed, so much so in one species that it is called the phosphorescent sea-pen (*Pennatula phosphorea*). Linnæus has stated that the phosphorescent sea-pens which cover the ocean bottom cast so strong a light that it is easy to count the fishes and worms which sport around them. Dr. Grant says that " a more singular and beautiful spectacle could scarcely be conceived than that of a deep purple pen, with all its delicate transparent polypi expanded, and emitting their usual brilliant phosphorescent light, sailing through the still and dark abyss, by the regular and synchronous pulsations of the minute fringed arms of the whole polypi." In the Scotch seas this species is found in great plenty, sticking to the baits on the fisherman's lines, especially when they make use of mussels to bait their hooks. Professor Edward Forbes conducted a series of experiments with these animals, in order to ascertain something of the nature of their phosphorescence, and from these he infers: (1) The polype is phosphorescent only when irritated by

FIG. 41.
POLYPES OF THE THORNY SEA-PEN.

P

touch; (2) The phosphorescence appears at the place touched, whether it be the stalk or the polyp-bearing part, and proceeds from thence in an undulating wave to the extremity of the polyp-bearing portion, and never in the other direction; (3) If the centre of the polyp-bearing portion be touched, only those polyps above the touched part give light; and if the extreme polyp-bearing pinna be touched it alone of the whole animal exhibits the phenomena of phosphorescence; (4) The light is emitted for a longer time from the point of injury, or pressure, than from the other luminous parts; (5) Sparks of light are sometimes sent out by the animal when pressed: these are found to arise from luminous matter investing ejected spicules. Subsequently the same observer writes that "unless the animal be in the highest state of vivacity, the stalk shows no phosphorescence, and the light of the feathered portion only runs a short way, but always towards the upper extremity. When plunged into fresh water the sea-pen scatters sparks in all directions,—a most beautiful sight,—but when plunged into spirits, it does not do so, but remains phosphorescent for some time, the light dying gradually away, and last of all from the uppermost polyps." A further number of experiments were made with the view of ascertaining whether the light was produced by electricity, and the results are given in Johnston's "Zoophytes" to the effect that electricity does not appear to be the cause, but, "on the whole, it is most probable that the animal secretes a spontaneously inflammable substance. It may be a compound of phosphorus, but it is not necessary to assume that it is."

There is a fascinating interest in the whole subject of phosphorescence in marine animals, and at the same time considerable mystery about its devolution. How it is produced, what its immediate use, under what conditions manifested, and whether always the same in its source and nature, are all problems constantly propounded, and never completely settled. Like the light itself, these questions constantly elude the grasp, and yet treatises have been written, and suggestions offered, time after time, which only partially, and very partially, meet the difficulty. Even as recently as the date of the "*Challenger* Expedition" the subject has come again to the surface, as the following allusions will indicate :—

While dredging in from 557 to 584 fathoms, Professor Wyville Thomson says:—" Many of the animals were most brilliantly phosphorescent, and we were afterwards even more struck by this phenomenon in our Northern cruise. In some places nearly everything brought up seemed to emit light, and the mud itself was perfectly full of luminous specks. The Alcyonarians, the brittle-stars, and some annelids were the most brilliant. The Pennatulæ, the Virgulariæ, and the Gorgoniæ shone with a lambent white light, so bright that it showed quite distinctly the hour on a watch; while the light from *Ophiocantha spinulosa* was of a brilliant green, coruscating from the centre of the disk, now along one arm, now along another, and sometimes vividly illuminating the whole outline of the star-fish" ("Depths of the Sea," p. 98). And again, whilst dredging near Shetland, he writes :—" Among Echinoderms *Ophiocantha spinulosa* was one of the prevailing forms, and we were greatly struck with the brilliancy of its phosphorescence. Some of these hauls were taken late in the evening, and the tangles were sprinkled over with stars of the most brilliant uranium green; little stars, for the phosphorescent light was much more vivid in the younger and smaller individuals. The light was not constant, nor continuous all over the star, but sometimes it struck out a line of fire all round the disk, flashing, or, one might rather say, glowing, up to the centre; then that would fade, and a defined patch, a centimètre or so long, break out in the middle of an arm, and travel slowly out to the point, or the whole five rays would light up at the ends, and spread the fire inwards. Very young Ophiocanthæ, only lately rid of their 'plutei,' shone very brightly. It is difficult to doubt that in a sea swarming with predaceous crustaceans, such as active species of *Dorynchus* and *Munida*, with great bright eyes, phosphorescence must be a fatal gift.

"We had another gorgeous display of luminosity during this cruise. Coming down the Sound of Skye from Loch Torridon, on our return, we dredged in about 100 fathoms, and the dredge came up tangled with the long pink stems of the singular sea-pen, *Pavonaria quadrangularis*. Every one of these was embraced and strangled by the twining arms of *Asteronyx loveni*, and the round soft bodies of the star-fishes hung from them like plump ripe fruit. The Pavonariæ were resplendent with a pale lilac phosphorescence, like the flame of cyanogen gas ; not scintillating, like the green light of *Ophiocantha*, but almost constant, sometimes flashing out at one point more brightly, and then dying gradually into comparative dimness, but always sufficiently bright to make every portion of a stem caught in the tangles, or sticking to the ropes, distinctly visible. From the number of specimens of Pavonaria brought up at one haul we had evidently passed over a forest of them. The stems were a mètre long, fringed with hundreds of polyps."[1]

On another occasion a speculation is indulged in, which may, or may not, be worthy of further consideration. Sir Wyville Thomson records a species

[1] "Depths of the Sea," p. 148.

of *Pleurotoma* which was dredged from 2,090 fathoms, and "had a pair of well-developed eyes on short footstalks; and a *Fusus*, from 1,207 fathoms, was similarly provided. The presence of organs of sight," he proceeds to say, "at these great depths leaves little room for doubt that light must reach even these abysses from some source. From many considerations it can scarcely be sunlight. I have already thrown out the suggestion that the whole of the light beyond a certain depth might be due to phosphorescence, which is certainly very general, particularly among the larvæ and young of deep-sea animals, but the question is one of extreme interest and difficulty, and will require careful investigation." [1]

As long since as 1853, M. de Quatrefages wrote a memoir on the phosphorescence of the lower marine animals, which may even now be consulted with interest,[2] in which he alludes to the fact that up to that date four hundred and fifty authors had treated, more or less fully, of the production of light by organised beings. Even an abstract of this memoir, which is in itself an abstract of antecedent observations, would scarcely be warranted here, and is scarcely necessary, as it is available in an English translation.

[1] "Depths of the Sea," p. 466.
[2] "Annals of Natural History," second series, vol. xii. pp. 16, 180.

CHAPTER VII.

CORAL BUILDERS.

WRITING concerning coral islands, and their inhabitants, a justly celebrated American Professor waxes eloquent over the errors of some of his predecessors, and even sometimes to the extent of being unduly hard upon their failings. "Many of those," he says, "who have discoursed most poetically on zoophytes have imagined that the polyps were mechanical workers, heaping up the piles of coral rock by their united labours; and science is hardly yet rid of terms which imply that each coral is the constructed hive or house of a swarm of polyps, like the honeycomb of the bee, or the hillock of a colony of ants." Although we feel a little sympathy with him in his indignation, it is doubtful whether he has not accepted in too literal a sense what he has characterised as a poetical discourse. Whether or not he has done so, it may somewhat mitigate his indignation if at once we accept his explanation of the process of manufacture.

"It is," he says, "not more surprising, nor a matter of more difficult comprehension, that a polyp should form structures of stone (carbonate of lime), called coral, than that the quadruped should form its bones, or the mollusk its shell. The processes are similar, and so is the result. In each case it is a simple animal secretion, a secretion of stony matter from the aliment which the animal receives, produced by the parts of the animal fitted for this secreting process; and in each carbonate of lime is a constituent, or one of the constituents, of the secretion. Coral is never, therefore, the handiwork of the many-armed polyps; for it is no more the result of labour than bone-making in ourselves. And again, it is not a collection of cells into which the coral animals may withdraw for concealment, any more than the skeleton of a dog is its house or cell; for every part of the coral of a polyp, in most reef-making species, is enclosed within the polyp, where it was formed by the secreting process." Great as may be our sin in calling the coral animals either "coral architects" or "coral builders," having thus limited the application of these terms, we may have some hope of escape. Furthermore, it is to be regretted that he should have applied the canon of scientific criticism to a poem with which he fails to find himself in sympathy, because it is a poetical work, and not a scientific treatise. "More error," he says, "in the same compass could scarcely be found than in the part of Montgomery's 'Pelican Island' relating to coral formations. The poetry of this excellent author is good, but the facts nearly all errors,—if literature allows of such an incongruity. There is no 'toil,' no 'skill,' no 'dwelling,' no 'sepulchre,' in the coral plantation, any more than in a flower-garden; and as little are the coral polyps shapeless worms that 'writhe and shrink their tortuous bodies to grotesque dimensions.' The poet oversteps his licence, and, besides, degrades his subject, when downright false to nature."[1] We might add to this the Professor's own words :—"It is not, perhaps, within the sphere of science to criticise the poet"; and evidently no one will think the less highly of a beautiful poem, as a poem, because it is by no means rigidly scientific, and even "the facts are nearly all errors." It had for its object the inculcation of truth, by means of fable, and not the dissemination of scientific information, on the structure and development of coral formations. We fear that we shall still remain heretical enough to delight in the "Pelican Island."

These preliminary observations introduce us at once to the subject of "coral builders," those minute marine animals which have done so much in the past, and are still at work in the present, in the con-

---

[1] Dana, "Corals and Coral Islands," p. 3.

struction of coral rock (fig. 42). Let those who please quibble over the terms of definition, and call it work or play, as suits them best; the animals are, nevertheless, the constructors of the coral rock. It is not so

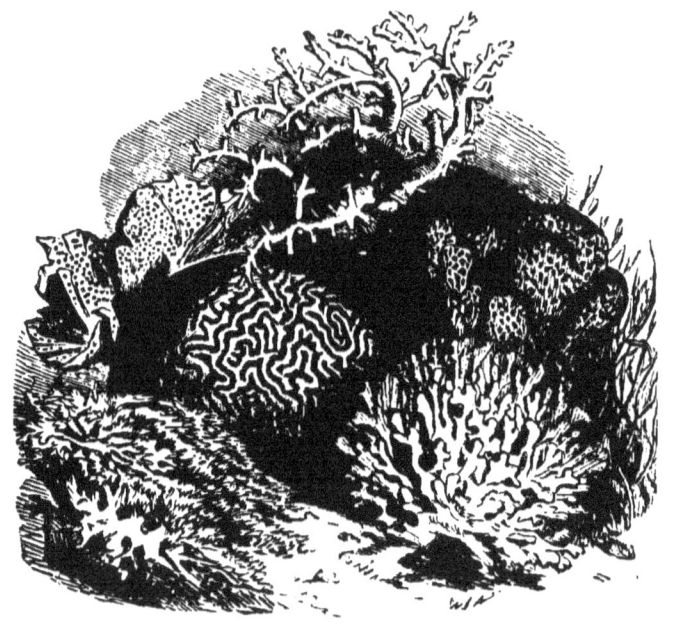

FIG. 42.—MADREPORES AND CORALS.

readily that we can select a type of a definite character, and, having briefly sketched its life history, point to it as the coral animal, as could be done with some other groups, since in this case there is more than one type of coral animal, and the details which would answer for one could not be applied to all. Hence it will be necessary to indicate, separately, the

Polyps, properly so called, and the Hydroids, commencing with the former.

If we select for illustration such a familiar marine animal as the sea-anemone, with which most persons are more or less acquainted in these latter days, either adhering to rocks, on the sea-shore, at low water, or flourishing in tanks, either in public aquaria, exhibitions, or zoological gardens, we shall have, as far as structure goes, a very fair representative of the Actinia-like coral polyps, but deficient in the power of secreting coral. With this exception, it will suit our purpose. Who has not read the Rev. Charles Kingsley's "Glaucus; or, Wonders of the Shore"? And whoever has read it will remember what he writes of these Anemones:—" In the crannies of every rock you will find sea-anemones. There they hang upon the under-side of the ledges, apparently mere rounded lumps of jelly; one is of dark purple dotted with green; another, of a rich chocolate; another, of a delicate olive; another, sienna-yellow; another, all but white. Take them from their rock; you can do it easily by slipping under them your finger-nail, or the edge of a pewter spoon. When you get home, turn them into a dish full of water, and leave them for the night, and go to look at them to-morrow. What a change! The dull lumps of jelly have taken root, and flowered during the night, and your dish is filled, from side to side, with a bouquet of chrysanthemums; each has expanded into a hundred-petalled flower, crimson, pink, purple, or orange; touch one, and it shrinks together like a sensitive plant, displaying at the root of the petals a ring of brilliant turquoise beads (fig. 43). That is the commonest of all the Actiniæ (*Mesembryanthemum*). You may have him when and where you will; but if you will search these rocks somewhat closer, you will find even more gorgeous species than he. See in that pool some dozen noble ones in full bloom, and quite six inches across, some of them. If their cousins, whom we found just now, were like chrysanthemums, these are like quilled dahlias. Their arms are stouter and shorter in proportion than those of the last species, but their colour is equally brilliant. One is a brilliant blood red; another, a delicate sea-blue, striped with pink; but most have the disk and the innumerable arms striped and ringed with various shades of grey and brown."[1]

FIG. 43.—SEA ANEMONE.

---

[1] "Glaucus; or, the Wonders of the Shore," by Rev. C Kingsley, M.A., p. 183.

Any one of these sea-anemones consists of a cylindrical body, or column, with a flattened base by which it is attached to the rock, a flattened upper surface or disk, surrounded by one or more rows of tentacles, and a mouth or opening in the centre, leading into the cavity of the body, or stomach, if you will. They may vary in size, from an eighth of an inch in diameter to more than a foot, but these general features will apply to all. When fully expanded the tentacles radiate from, and fringe the disk, like the flower of a chrysanthemum. When retracted and closed, the tentacles are hidden, and the whole animal resembles a convex gelatinous button. The number of tentacles is some multiple of six, which is also a typical number in their internal organisation. The mouth, in the centre of the disk, is usually a little elevated at the margin or lips, and opens directly into the stomach. This cavity is divided, in a radiate manner, by fleshy partitions, or septa, into compartments, the septa being in pairs, corresponding to some multiple of six. The muscles of the body are the means by which the animal contracts itself when disturbed, and expands again when at rest; during the former process expelling forcibly a quantity of water through its mouth, and in the latter imbibing sufficient to inflate the tentacles and body. Food is sometimes aśsisted in its passage through the mouth into the stomach by the tentacles, as well as by the extension of the lips, and the contraction of the disk. In some species the tentacles are too short and thick to be of much service for prehension. When the food is digested, or as much of it as is capable of digestion, the refuse is expelled from the mouth, which is the only orifice of ingress and egress.[1] "When morsels of food," says Gosse, "such as fragments of butcher's meat, are swallowed by Anemones, they are retained for some hours and then vomited ; and, because little change has passed upon the solid parts, it has been rashly concluded that no process of digestion takes place in these animals. On this foolish hypothesis it is difficult to see why food should be swallowed at all, or what need the animal has of mouth or stomach. Their ordinary food, however, is not mammalian muscle, but far softer and more fluid flesh. Nothing is more common than to find large specimens of Actiniæ discharge soon after their capture, the shell of a crab or limpet from which the entire flesh has been removed, and replaced by a tenacious glaire. No doubt the first part of the process consists largely of maceration, and continued pressure, by means of which the juices of the food are extracted." Without entering further upon the

---

[1] All the details of structure and economy of these animals will be found in the Introductory Chapter to "A History of Sea-Anemones," by P. H. Gosse, London, 1860.

question of digestion, or of circulation and respiration, a few lines must be devoted to the powers, and means of attack and defence, possessed by these apparently harmless animals.

The concealed weapons of the sea-anemones, and their allies, consist in the lasso-cells, stinging-cells, or thread-capsules, as they have been called, which are distributed over the animals in myriads. Gosse has described minutely the different forms, or modifications, of these lasso-cells, but his description is too extended and technical for repetition here (fig. 44). Suffice it to say that, according to Dana, they " are called *lasso-cells* because the little cell-shaped sheath contains a very long slender tubular thread coiled up, which can be darted out instantly when needed. As first observed by Agassiz, the tubular lasso escapes from the cell by turning itself inside out, the extremity showing itself last, and this is usually done with lightning-like rapidity. Then follows the poison. The lasso-cells are usually less than one two-hundredth of an inch in length, but they are

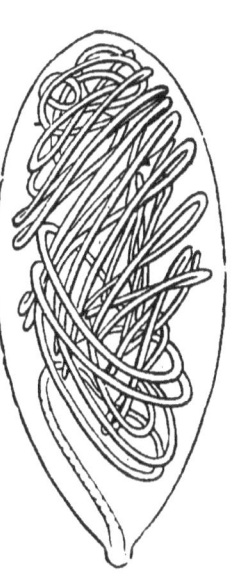

FIG. 44.
LASSO-CELLS OF CORYNACTIS.

thickly crowded in the larger part of the skin or walls of the tentacles and about the mouth, also in the visceral cavity in white cords hanging in folds from the edge of the septa. Thus the polyp is armed inside and out. The mollusk or crab that has the ill luck to fall, or be thrown by the waves, on the surface of the pretty flower is at once pierced and poisoned by the minute lassoes, and is rendered incapable of resistance." [1]

In the particular form of lasso-cell, here figured

FIG. 45.—LASSO-CELL OF MADREPORE.

(fig. 45), the cells are perfectly transparent, colourless vesicles, of a long ovate figure, larger at one end than the other. One of average size is one two-hundred-and-fiftieth of an inch in length, and about one two-thousandth of an inch in diameter, at its widest part. In the upper and larger half of the cell a longitudinal chamber of a spindle-shape is seen passing downwards through its centre, tapering to a point at each end; the upper point merges into the walls of the cell at its extremity, whilst the lower point, after

---

[1] Dana, " Corals and Coral Islands," p. 13.

being narrowed like the upper, expands into a funnel-shaped mouth in which the chamber wall is folded inwards. From this point the structure becomes a slender cord, which, passing back to the upper end of the capsule, winds loosely round and round the chamber, at first regularly, but afterwards in a more intricate manner, until it fills the lower portion of the cavity. Under pressure, or at the will of the animal, the fusiform chamber and its twining thread are shot forth with inconceivable rapidity. When fully expelled the thread is often twenty, thirty, or even forty times the length of the lasso-cell, though in some species it is much shorter. It may be added that the basal portion of the thread (lasso) is often ornamented by a thickened spiral band, in which are inserted a series of firm tapering bristles (fig. 46).

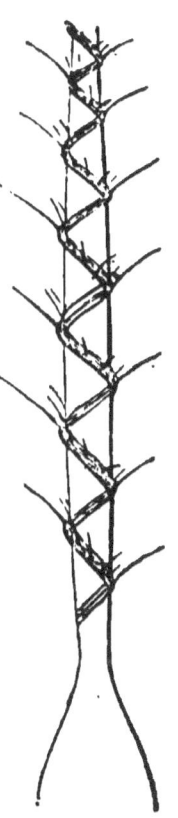

FIG. 46.
BASAL PORTION
OF LASSO.

"It has long been known," says Gosse, "that a very slight contact with the tentacles of a polype is sufficient to produce, in any minute animal so touched, torpor and speedy death. Since the dis-

-covery of these lasso-cells the fatal power has been supposed to be lodged in them. Baker, a century ago, in speaking of the Hydra, suggested that 'there must be something eminently poisonous in its grasp,' and this suspicion received confirmation from the circumstance that the *Entomostraca* (water-fleas, &c.), which are enveloped in a shelly covering, frequently escape unhurt after having been seized. The stinging power possessed by many *Medusæ*, which is sufficiently intense to be formidable even to man, has been reasonably attributed to the same organs, which the microscope shows to be accumulated by millions in their tissues."[1] This author goes on to say that, although he cannot reduce this presumption to actual certainty, he made some experiments which leave no reasonable doubt on the subject. He then proceeds to record several examples, which seemed to prove that the slightest contact with the proper organs of the anemone was sufficient to provoke the discharge of the lasso-cells, and that even the densest condition of the human skin offered no impediment to the penetration of the threads. And he adds,—" As to the injection of a poison, it is indubitable that pain, and in some cases death, ensues, even to vertebrate animals, from momentary contact with the capsuliferous organs of the zoophytes. The very severe

---

[1] Gosse, " Sea-Anemones and Corals," p. xxxvii.

pain, followed by torpor, lasting a whole day, which Mr. George Bennett has described as experienced by himself on taking hold of a 'Portuguese man-of-war' (*Physalis pelagica*) was produced by the contact of the tentacles. The late Professor Edward Forbes has graphically depicted the 'prickly torture' which results to 'tender-skinned bathers' from the touch of the long filamentous tentacles, 'poisonous threads' of the *Cyanea capillata* of our own seas, and observes that these amputated weapons, severed from the parent body, sting as fiercely as if their original proprietor itself gave the word of attack. I have been assured by ladies that they have felt a distinct stinging sensation, like that produced by the leaves of the nettle on the tender skin of the fingers, from handling our common 'Opelet anemone' (*Anthea cereus*) ; while, on the other hand, I have myself handled the species scores of times with impunity. And I have elsewhere recorded an instance in which a little fish, swimming about in health and vigour, died in a few minutes with great agony, through the momentary contact of its lip with one of the emitted lasso-cells (*acontia*) of the 'Parasitic anemone' (*Sagartia parasitica*). It is worthy of observation that in this case the fish carried away a portion of the lasso-cell (*acontium*) sticking to its lip ; the force with which it adhered being so great that the integrity of the tissues yielded first. The lasso (*acontium*) severed rather than let go its hold".[1] (fig. 47).

The anemone has no circulating fluid but the results of digestion mixed with salt water, no blood-vessels but the vacuities among the tissues, and no passage way for excrement, except the mouth and

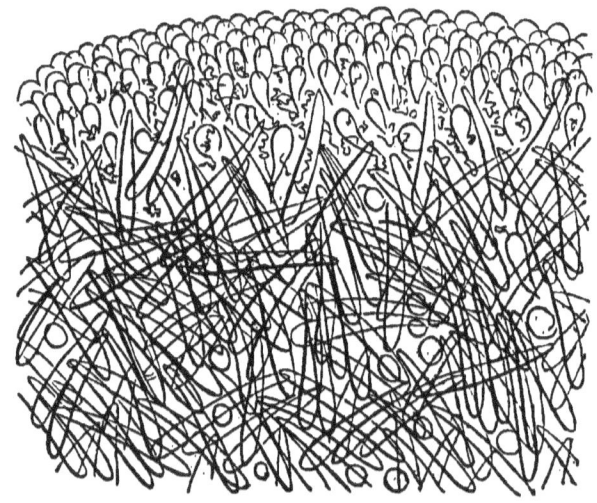

FIG. 47.—ACONTIA OF SEA-ANEMONE.

the pores of the body, that serve for the escape of water on the contraction of the animal.

---

[1] Mr. G. H. Lewes, in his "Seaside Studies," doubts the validity of this fact, and indeed he claims that "the hypothesis which assigns to the thread capsules a function of urtication is without a single fact to warrant it."

As to the senses they have a general sense of feeling, and besides which, "some of them have a series of eyes, placed like a necklace around the body, just outside the tentacles," yet they are not known to have a proper nervous system.

Reproduction takes place either by means of eggs or by budding. Sometimes an animal divides itself from above downwards, and becomes two individuals, or fragments detached from the base will develop themselves into young individuals, but gemmation, or budding, is a common process, especially amongst the coral-secreting species. "A protuberance begins to rise," says Dana, "and soon shows a mouth, and then becomes surrounded by tentacles; and, thus begun, the new anemone continues to grow, usually until its tentacles have doubled their number, when finally it separates from the parent, and becomes an independent animal."

The normal mode of reproduction is by generation. The sexes are sometimes united in one individual, and sometimes separate. The eggs, germs, or fully-formed young, are discharged indifferently through the mouth: in the latter case, the embryos have passed their early stages in the general cavity. Dr. Spencer Cobbold relates how he collected two specimens of anemone one afternoon from the north shore of the Frith of Forth : " On arriving at Edinburgh, the same evening, the number in the vessel had increased to thirty-five. The animals were therefore carefully watched that night, but not until next day were we gratified by witnessing, what has been shown by numerous observers to occur, viz., the evolution of the young, of all shapes and sizes, by the mouth. On dissecting one of the adult anemones, it was found, as we were thus led to anticipate, that a considerable opening obtained to the base of the stomach, admitting the tip of the little finger, and freely communicating with the interseptal spaces and general abdominal reservoir. This, to our mind fully cleared up the difficulty, so often expressed, concerning the passage by which the young polyps gain access to the digestive cavity, and are ultimately set free."[1]

Thus far then we have detailed the structure and development of the common anemones, which do *not* secrete coral, in order to illustrate the coral-making species. The differences between them consist in the anemones being simple animals,—that is, not associated to form colonies,—in their being generally capable of locomotion on the muscular base, as well as in the deficiency of the power of secreting coral. In other respects, that which applies to the one group applies also to the other. In consequence of their

---

[1] Dr. S. Cobbold on the "Anatomy of Actinia," in "Annals and Magazine of Natural History." Feb. 1853, p. 122.

depositing within themselves a skeleton of coral, the coral-making species are fixed (fig. 48), this deposit taking place in the sides, and lower part of the polyp, leaving the disk and stomach fleshy. The internal septa secrete radiating plates of coral between each pair of septa, so that when dead, and the fleshy parts are dissolved away, these radiating plates represent the radiating septa in the internal cavity of the anemone. The calcareous, or persistent coral skeleton, which is left when all the fleshy parts are gone is called the *corallum*, and is, in fact, the coral.

FIG. 48.—CARYOPHYLLIA SMITHII.

A few of the species in this group are simple, and the coral produced is the secretion of single polyps. This is the case in the Fungia family in which the numerous radiating plates resemble the gills of a mushroom inverted, whence their name is derived. They are by no means small species, but will attain to more than a foot in length, the corallum, or skeleton, being a common object in museums. They are found in coral-reef seas lying on the bottom, but, unlike most coral-secreting species, are not fixed except when young. Their outline is often discoid or oval, but sometimes elongated elliptical, with the plates on the upper surface radiating from a depressed umbilicate centre, or from a central furrow, the outer edges of the parallel plates being acute.

The most important of the Actinoid [1] corals are those which form compound groups, wherein a number of polyps are connected together in a colony, often as the result of the process of budding. These colonies Professor Dana calls *Zoothomes*, although we prefer to avoid as much as possible the use of technicalities of this description. Describing the process of gemmation, or budding, he says, " the bud commences as a slight prominence on the side of the parent. The prominence enlarges, a mouth opens, a circle of tentacles grows out around it, and increase continues till the young family equals the parent in size. Since in these species the young does not separate from the parent, this budding produces a compound group ; and the process continues until, in some instances, thousands, or hundreds of thousands, have proceeded from a single germ, and the colony has increased to a large size, sometimes many feet, or even yards, in breadth or height. In the corallum, or dead skeleton, each stellate cavity, or prominence

---

[1] Actinoid, because resembling the Actiniæ, or Sea-anemones.

CORAL BUILDERS. 233

over the surface, corresponds to a separate one of the united polyps." Each individual polyp has its own separate mouth, tentacles, and stomach, and they coalesce by intervening tissues, so that " there is a free circulation of fluids through the many pores, or lacunæ. The colony is like a living sheet of animal matter, fed and nourished by numerous mouths, and as many stomachs." The budding proceeds in different directions according to the species. In some cases the base spreads in all directions, and buds at the edge, so as to form an incrusting plate, in other cases it has a tendency to grow upwards, budding at the sides, so as to form branches, and at length a tree-like form is produced, or the budding takes place all over the mass, so that it maintains a more or less hemispherical form, some of the masses in the Pacific having a diameter of twenty feet. Thus we have the numerous forms of *Orbicella*, *Madrepores*, *Porites*, &c.

But, besides this method of budding, there is another and analogous process of spontaneous fission by which a polyp divides in two. In such cases the disk keeps enlarging until, having exceeded the ordinary adult size, a new mouth makes its appearance in the disk, a short distance from the old one, growth still continues until the mouths are removed further apart, a stomach is being formed beneath the new mouth, and at length tentacles make their appearance between the two mouths, each mouth with its circlet of tentacles, and finally these are resolved into two complete individuals, the one retaining the old mouth and stomach, the other the new mouth and stomach. This is the mode of increase which commonly prevails in the hemi-

FIG. 49.—THE ASTRÆA (*Astræa pallida*).

spherical corals of the genus *Astræa* (fig. 49), some of which attain a diameter of from ten to fifteen feet.

A curious modification of this process takes place in the Brain corals (*Meandrina*), in which the disk enlarges indefinitely, a new mouth makes its appearance, and then another, and another, until a string of mouths appear in a line, before there is any separation or isolation of the new individuals; this, however, does take place at length, preparatory to the formation by each of its own elongating disk and line of new mouths. This serves to explain the

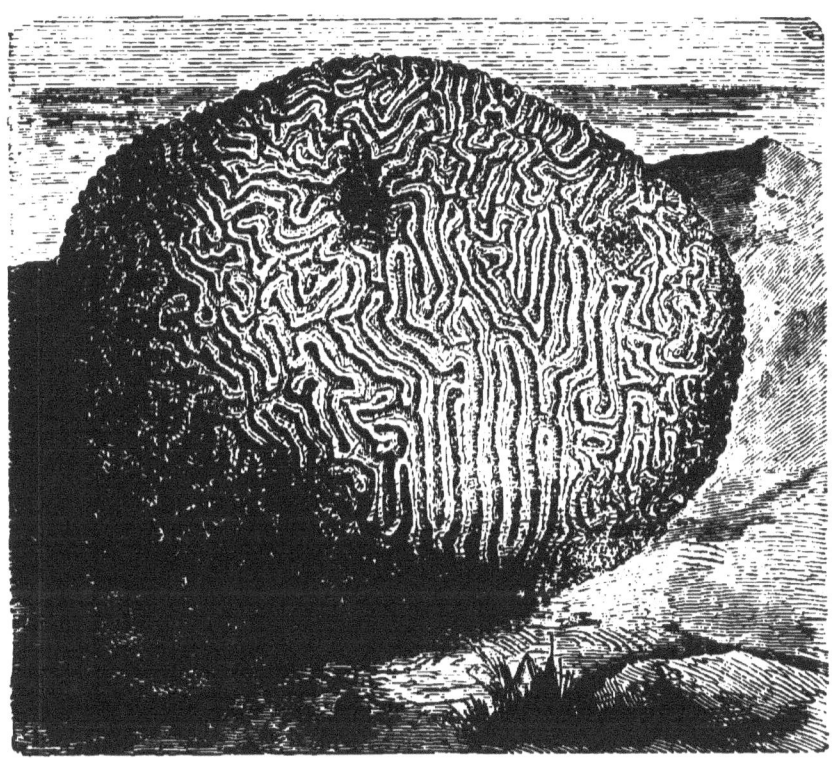

FIG. 50.—BRAIN CORAL (*Meandrina cerebriformis*).

sinuous furrows on the surface of the corallum of the Brain corals (fig. 50).

The only other important group of coral animals to which it will be necessary to allude, are the Hydroids; and here we may fall back upon what we have already written for many of the facts, to avoid repetition. The hydroid zoophytes, as they are often called, will hereafter be described, inasmuch as concerns the "Horny House-builders." These may be taken to a great extent as a type of the Hydroid coral animals, since, in both, the common fresh-water Hydra furnishes a good illustration of general structure. When speaking or writing of corals, allusion is often made to a group called Millepores, in which the structure of the animals recedes from that of the Actinoid corals, to which our space has hitherto been devoted. These animals are slow to make their appearance; and, indeed, it has not long been known to what precise position in the animal kingdom they should be assigned. At length, however, the American naturalist, Agassiz, determined their character, which seems to be generally accepted, that they are Hydroids, allied to our own *Coryne*, but less perfectly developed. The animal appears to be only a fleshy tube, with a mouth at the top, and, in place of the tentacles, a few small rounded prominences, four of which are somewhat the largest. The corals are large and stony, punctured all over the surface with small rounded holes, through which the animals protrude themselves, so that the corals themselves may be readily distinguished from those of the anemone type. As these corals are very abundant, and contribute largely to the formation of reefs in the West Indies, they could not be ignored in this chapter, although their interest is small, as compared with the various forms of Actinoid corals.

It must, in justice, be remarked here that the conclusions to which Agassiz arrived have not remained unchallenged. Observations on the same species as that examined by the American zoologist, made by another individual, have led to the adoption of different views, which are thus stated : " The polyp, as seen at Bermuda in full expansion, is a very remarkable one, and it is a great satisfaction to be able to state that Agassiz saw only a part of the whole, and came to his conclusions too rapidly. The polyps are of different lengths, according to their growth, are slender, and stand erect in crowds around the branches.

FIG. 51.
ZOOIDS OF MILLEPORA.

Each rises from a cylindrical stem, which is rendered slightly square close to four tentacles, which project upwards and outwards. Their tips are swollen and rounded, and their bases are continuous, by means of straight dark tissue, which overlaps slightly the analogue of the oral opening (fig. 51). Out of this opening comes a second cylinder, to terminate in four other tentacles in the same way; and in some polyps there is a further growth, so that there are two or more rows of tentacles separated by the tubular cylindrical tissue. It is evident that Agassiz saw young, ill-developed, and probably injured polyps, which had not attained their second row of tentacles. The number of tentacles may be, therefore, four, eight, twelve, &c. In looking at this description there is a probability that the Millepore is an Alcyonarian, and there is no proof that it is a Hydroid."[1] It is also stated that "every variety of tentacular and disk apparatus may exist in either; but the external development of the gemmules, ova, and embryonic forms, must be recognised before any Cœlenterate animal can be associated with the Hydrozoa. Here is the point where Agassiz fails. His researches are only suggestive, until the generative organs are recognised on the protruded polyps."

In his enumeration of the different groups of animals which contribute to the formation of coral reefs, Dana includes the Sea-fans, and their allies, and the Sea-mat animals (*Bryozoa*). The former of these he admits contribute but little to the material

---

[1] "On the Actinozoan Nature of Millepora," by R. G. Nelson, and P. M. Duncan, in "Annals of Natural History," vol. xvii. (1876), p. 354.

of coral-reefs, but they add largely to the beauties of the coral landscape. The animal of the commercial Red Coral belongs to the same category, being what is termed an Alcyonoid polyp, and all further allusion to it will be found in our chapter on Sea-fans.

The Bryozoans, or animals of the sea-mats, are of another and distinct type, but the structures they form are usually small, and either thin and membranaceous, or thicker and more calcareous, forming plates which may, by growing in a stratified manner one over the other, ultimately produce a layer of considerable thickness. Whatever they may have done, in Palæozoic times, in contributing to the formation of limestone, they contribute but little to the coral-reefs of the present age. In furtherance of our design, if space permits, we hope to include both these groups of animals in the present volume, therefore are content to abandon them for the present.

It is also a fact that some undoubted plants, or sea-weeds, secrete lime to such an extent as to contribute to the accumulations which form coral-reefs. Some of these are very delicate and beautiful, though fragile, and their name of *corallines* suggests miniature resemblances to coral. The larger kinds, or Nullipores, not only contribute something in bulk to the calcareous deposit of reefs, but Darwin says that "they play an important part in protecting the upper surfaces of coral-reefs, behind, and within, the breakers, where the true corals, during their outward growth, become killed by exposure to the sun and air."

The corallum, or coral skeleton, which is secreted by the coral animals, and enters into the composition of coral-reefs and coral-rock, is almost identical with ordinary limestone. It is somewhat harder than limestone or marble, and rings when struck with a hammer. "It is a common error," writes Dana, "to suppose that coral, when first removed from the water, is soft, and afterwards hardens on exposure; for, in fact, there is scarcely an appreciable difference. The live coral may have a slimy feel in the fingers, but, if washed clean of the animal matter, it is found to be quite firm. The water with which it is penetrated may contain a trace of lime in solution, which evaporates on drying, and adds slightly to the strength of the coral, but the change is hardly appreciable." Chemically, the coral-reefs are composed of carbonate of lime, viz., they contain from ninety-five to ninety-eight parts in one hundred of that constituent,[1] and the coral animals are the medium, through which this is abstracted from the sea water, and deposited in

---

[1] Madrepora cervicornis, contains in 100 parts 98·07 of carbonate of lime, 0·32 phosphate of lime, and 1·93 of water and organic matters.—*American Journ. Sci.*, iii. i. 168.

their skeleton. It is calculated that the water of the ocean contains, in solution, from one twenty-fourth to one thirty-sixth of all its soluble ingredients in salts of lime, chiefly in the condition of sulphate of lime; and, according to Bischof, there are about sixteen parts of sulphate of lime in 10,000 of sea water. This proportion may at first appear to be small, but by rapid infiltration, and deposit, by myriads of animals, it has been considered ample to account for the results. After all, no one can estimate the enormous period of time throughout which the coral has been deposited to form existing reefs. This leads us to an interesting subject, suggested by Professor Dana, in his work already so often quoted, and this is, death and life in progress together in the coral structure. No one must suppose that a mass of coral, which may be dredged up from the sea-bottom, is necessarily living from end to end. Whatever the species may be, and in whatever direction its growth may proceed, there will be found to be an old and dead portion, with a new and vigorous one,—death and life side by side, and both going on together. The animal portion decays and vanishes, but the corallum, or coral stock, remains imperishable for ages. It is by means of this process that masses of coral, which would otherwise be very small, attain to a very large size. The animals at the top, as they deposit their coral, keep ascending, whilst the older polyps below die and disappear. " Trees of Madrepores have their limits,—all below a certain distance from the summit being dead, and this distance will differ for different species. But this is not a limit to the existence of the corallum, even though a slender tree or shrub, or of its flourishing state ; for the dead coral below is firm rock itself, often stronger than ordinary limestone or marble, and serves as an everrising basement for the still expanding and rising zoophyte." Not only will the base be dead, and the apex of the trunk or branches living, but the whole interior is usually dead, so that the living portion does not extend inwards but the fraction of an inch. In the large dome-shaped corals, sometimes 10 feet or 15 feet in diameter, the whole outer surface may be alive, and yet the whole interior be nothing but lifeless coral. The tree-like species may continue to increase upwards, until the apex reaches the surface at low water, and then cease, for death comes from exposure, and yet not wholly cease ; for even under such circumstances lateral budding will still go on with a modification of form.

The dead coral trunks are not without their enemies, and their great hardness does not secure them wholly from danger. Agassiz states that "innumerable boring animals establish themselves in the lifeless stem, piercing holes in all directions into its interior, like so many augers, dissolving its solid connexion with the ground, and even penetrating far into the living portion of these compact communities. The number of these boring animals is quite incredible, and they belong to different families of the animal kingdom; among the most active and powerful we would mention the date-fish, and many worms, of which the Serpula is the largest and most destructive, inasmuch as it extends constantly through the living part of the coral stems. On the loose basis of a brain coral (*Meandrina*), measuring less than two feet in diameter, we have counted not less than fifty holes of the date-fish—some large enough to admit a finger —beside hundreds of small ones made by worms. But however efficient these boring animals may be, in preparing the coral stems for decay, there is yet another agent, perhaps still more destructive, we allude to the minute boring-sponges, which penetrate them in all directions, until they appear at last completely rotten through."[1] Yet the coral animals would seem to contribute themselves a little protection from their enemies, for in some of the species the lower edge of the dying polyps secrete an outer layer of denser, and more impervious, material over the dead coral stem, and, besides this, the older polyps before death increase their coral secretions within, so as to fill up gradually the pores of the coral, as their own

---

[1] Agassiz. Coast Survey Report for 1851.

tissues dwindle, and so render the old coral more nearly solid. Then again, some little external assistance is rendered by the mollusks, which adhere to the dead trunk, as well as by the Nullipores which grow over it, like the lichens on the trunk of an old tree in a forest. Notwithstanding all their vicissitudes in this "struggle for existence" the coral masses, and coral reefs, must for ages have maintained their position, in defiance of all their enemies, even allowing for all loss by disintegration, a process so continuous, and yet so small in its results, in comparison with the permanent mass, that it can scarcely be taken into account as a disturbing influence.

As we commenced this chapter by observations on some strictures which had been made on certain authors, on the grounds of scientific accuracy, so we feel bound to conclude it by some strictures of our own on the same grounds. Although it may be permissible for a poet to use what machinery he pleases, in the construction of his work, whether based wholly, or only in part, on scientific fact, it is nevertheless imperative on one who writes in a scientific work, assumed to give instruction in science, that he should be accurate in his interpretations of the facts of science. Whether this demand has been acceded to in the work in question the following quotation may determine. Writing of the coral animals this writer observes :—" Nothing can be more impressive than the manner in which these diminutive creatures carry out their stupendous undertakings, which we denominate instinct, intelligence, or design. Commencing betimes from a depth of a thousand or fifteen hundred feet, they work upwards in a perpendicular direction; and on arriving at the surface form a crescent, presenting the back of the arch in that direction from which the storms and winds generally proceed; by which means the wall protects the busy millions at work beneath and within. These breakwaters will resist more powerful seas than if formed of granite; rising as they do in a mighty expanse of water, exposed to the utmost powers of the heavy and tumultuous billows that eternally lash against them. As we glance at the map of the world, and think of the profusion of fragrant vegetation, and delicious food, almost spontaneously produced on the lovely sunny islands of the Pacific, how startling does it seem when we are told that these islands, bearing on their bosoms gardens of Eden, are entirely formed by the slow-growing corals, which, rising up in beautiful and delicate forms, displace the mighty ocean, defy its gigantic strength, and display a shelly bosom to the expanse of day! The vegetation of the sea, cast on its surface, undergoes a chemical change; the deposit from rains aids in filling up the little gaping catacomb, the fowls of the air and ocean find a resting-place, and assist in clothing the rocks; mosses carpet the surface, seed brought by birds, plants carried by the oceanic currents, animalcules floating in the atmosphere, live, propagate and die, and are succeeded, by the assistance their remains bestow, by more advanced vegetable and animal life; and thus generation after generation exist and perish, until at length the coral island becomes a paradise, filled with the choicest exotics, the most beautiful birds, and delicious fruits, amongst which man may indolently revel to the utmost desire of his heart."[1]

No doubt its author considered this a sublime finish to his meagre statement of facts, but, unfortunately it must have been educed from his own inner consciousness, and should be treated as the lucubration of a heated imagination, and not as a summary of scientific fact such as may be gleaned from the works of Darwin, Dana, Agassiz, and others, who have made the subject a special study. "How startling does it seem when we are told" that these gorgeous dreams are the stern realities of scientific fact, but perhaps the writer himself was scarcely conscious, in his enthusiasm, that his spirit was wandering in the unsubstantial realms of Fairy-land.

---

[1] "The Microscope," by Jabez Hogg, F.L.S. (London, 1867), p. 490.

CHAPTER VIII.

CORAL REEFS, AND ISLANDS.

THE first author who collated the information associated with Coral Islands, and propounded a feasible theory of their structure and method of construction, was the late Charles Darwin, who has done so much for Biological Science during the present generation. His views, propounded in full in his volume on Coral Reefs,[1] were subsequently subscribed to by the eminent American geologist, Professor James Dana,[2] and have been generally accepted; such exception as may have been taken to them being alluded to hereafter.

Darwin classed all the phenomena of coral reefs under three primary classes, Barrier reefs, Fringing reefs, and Atolls, or lagoon islands. These may be briefly described, in Mr. Darwin's own language, as

---

[1] "Structure and Distribution of Coral Reefs," by Charles Darwin. London, 1851.
[2] "Corals and Coral Islands," by James D. Dana, LL.D. Popular edition. London, 1875.

he defined them in his "Journal":—"Barrier reefs either extend in straight lines in front of the shores of a continent, or of a large island, or they encircle smaller islands; in both cases being separated from the land by a broad, and rather deep, channel of water." Fringing reefs, where the land slopes abruptly under water, are only a few yards in width, forming a mere ribbon or fringe round the shores; where the land slopes gently under water the reef extends further, sometimes even as much as a mile from the land; but in such cases the soundings outside the reef always show that the submarine prolongation of the land is gently inclined. As far as the actual reef of coral is concerned, there is not the smallest difference in general size, outline, grouping, and even in quite trifling details of structure, between a barrier reef and an atoll, or lagoon island (fig. 52). The geographer Balbi has well remarked, that an encircled island is an atoll with high land rising out of its lagoon; remove the land from within, and a perfect atoll is left."

These distinctions between coral reefs and coral islands (lagoon islands, or atolls, whichever name be employed) are recognised also by Dana, although he combines the two kinds of reefs, and generally treats of coral reefs (barrier reefs, and fringing reefs) and coral islands. "Coral reefs and coral islands," he says, "are structures of the same kind under somewhat different conditions. The terms, however, are not synonymous. *Coral islands* are reefs that stand isolated in the ocean, away from other lands, whether now raised only to the water's edge, and

FIG. 52.—CORAL REEF.

half submerged, or covered with vegetation; while the term *Coral reef*, although used for reefs of coral in general, is more especially applied to those which occur along the shores of high islands and continents."

With this preliminary definition of terms we are

enabled better to proceed with a more minute description of reefs and atolls, the latter term being less open to objection than islands.

Coral reefs are accumulations of coral rock which have been built up from the sea bottom around the shores of islands, or the continent, in the tropics. These reefs are usually entirely submerged at high water, and at low water present only a flat surface of rock just rising above the water level. Some islands are only girt by narrow fringing reefs, whilst others are wholly or partially surrounded by the distant barrier reef, which may be from ten to fifteen miles from the island, or islands, with a considerable channel of water between the barrier and the shore; and, in other cases, one side of the island may be protected by a distant barrier reef, and the other girt only by a fringing reef, the relationship between the barrier and fringing reef being shown to be very close. Hence we may formulate the conclusion that "reefs around islands may be entirely encircling, or they may be confined to a larger or a smaller portion of the coast, either continuous or interrupted; they may constitute throughout a distant barrier; or the reef may be fringing in one part, and a barrier in another; or it may be fringing alone. The barrier may be at a great distance from the shores, with a wide sea within, or it may so unite to the fringing reef that the channel between will hardly float a canoe."[1] Although reference has been made only to such portions of the reef formations as have been reared to the water's level, it must be remembered that there may also be submerged banks, which are in effect continuous with the elevated portions, and form a part of the reef ground. The extent to which these structures can be developed may be inferred from the dimensions recorded; as, for instance, to the west of the two large Fiji Islands there is probably three thousand square miles of reef-ground. The Exploring Isles have a barrier eighty miles in circuit; the western shore of New Caledonia has a reef throughout its extent of two hundred and fifty miles, and for a hundred and fifty miles beyond; whilst the Australian barrier forms a broken line of some one thousand two hundred and fifty miles in length.

Not only are we interested in the area which may be covered by a reef-ground, but also of the thickness to which some of the reefs may attain, and here again we must fall back on the calculations made by Dr. Darwin and Professor Dana. If it were possible to raise one of these coral-girt islands we should find that the barrier reefs stand like walls upon the deeply-submerged slopes, or, in the case of the fringing reefs, in the shallowest water, approaching

---

[1] Dana, "Corals and Coral Islands," p. 105.

the coast line. Having found the declivity of the land, assuming the slope to be regular and continuous, it would not be difficult to estimate the depth, at any given distance from the land, but there must be uncertainty as to the submarine slope, although extensive observation has shown that, in general, these slopes nearly coincide with those of the exposed land. In this, or a similar manner, Dr. Darwin estimated that the thickness of some of the reefs in the Pacific islands, at their outer limits, would be at least two thousand feet. After estimating the outer reef of the Gambier group at eleven hundred and fifty feet, of Tahiti at two hundred and fifty feet, and of Upolu at four hundred and forty feet, Professor Dana remarks that "the results are sufficiently accurate to satisfy us of the great thickness of many barrier reefs ;" and again, "with regard to Tahiti and Upolu, information bearing upon this point was obtained, and the above conclusions may be received with much confidence. Many of the Fiji reefs, on the same principle, cannot be less than two thousand feet in thickness." The objection which will at once be urged against such a thickness of coral, on the ground that living coral can only be found at a limited depth below the surface, will be adverted to hereafter, and perhaps reconciled with these estimates. For the present we leave the data as they stand.

CORAL REEFS, AND ISLANDS. 253

Coming now to the atolls, or lagoon islands, we may repeat that they resemble in many points the encircling reefs just described, but differ in enclosing a lake or lagoon instead of mountainous islands. The lagoon is indeed only an enclosed portion of the sea, surrounded by a more or less complete wall or belt of coral rock, which is but little elevated above the surface, and here and there surmounted by vegetation, which have the appearance of small islets. "In many of the smaller atolls the lagoon has lost its ocean character, and become a shallow lake, and the green islets of the margin have coalesced in some instances into a continuous line of foliage. Traces may, perhaps, be still detected of the passage, or passages, over which the sea once communicated with the internal waters, though mostly concealed by the trees and shrubbery which have spread around and completed the belt of verdure (fig. 53). The coral island is now in its most finished state; the lake rests quietly within its circle of palms, hardly ruffled by the storms that madden the surrounding ocean."[1]

It must not be assumed that the atolls are always ring-shaped, although that may be the most perfect form, for they are not only triangular, quadrangular, and very irregular, but sometimes are completed only on one side, with the opposite so much submerged that

---

[1] Dana, p. 131.

no trace of it can be seen at the surface. It might be supposed that when these atolls are called "coral islands" they are really extensive surfaces of habitable land, but such is not the case, for the small amount of habitable land they present is one of

FIG. 53.—CORAL LAGOON.

their most peculiar features. Dana has shown that in the Kingsmill group of ten islands, which have an aggregate area of eighteen hundred and fifty square miles, the actual amount of dry and habitable land is only seventy-six miles, or less than one twenty-fourth. In the Caroline Archipelago the proportion is still smaller, and in the Marshall Islands the dry land is not more than the one-hundredth part of the whole.

Professor Dana has given so graphic a description of an atoll that we are tempted to quote it, rather than submit a more prosaic one of our own. "The reef of the coral atoll, as it lies at the surface, still uncovered with vegetation, is a platform of coral rock, usually two to four hundred yards wide, and situated so low as to be swept by the waves at high tide. The outer edge, directly exposed to the surf, is generally broken into points and jagged indentations, along which the waters of the resurging wave drive with great force. Though in the midst of the breakers the edge stands a few inches, and sometimes a foot, above other parts of the platform, the encrusting Nullipores cover it with varied tints, and afford protection from the abrading action of the waves. There are usually eighteen to thirty feet of water near the margin, and below, over the bottom, which gradually deepens outward, beds of coral are growing profusely among extensive patches of coral sand and fragments. Generally the barren areas much exceed those flourishing with zoophytes, and not unfrequently the clusters are scattered like tufts of vegetation in a sandy plain. The growing corals extend up the sloping edge of the reef nearly to low-tide level.

"The beach consists of coral pebbles, or sand with some worn shells, and occasionally the exuviæ of crabs and bones of fishes. Owing to its whiteness and the contrast it affords to the massy verdure above, it is a remarkable feature in the distant view of these islands, and often seemed like an artificial wall or embankment running parallel with the shores.

"The emerged land beyond the beach, in its earliest stage, when barely raised above the tides, appears like a vast field of ruins. Angular masses of coral rock, varying in dimensions from one to a hundred cubic feet, lie piled together in the utmost confusion, blackened by exposure or from encrusting lichens. Such regions may be traversed by leaping from block to block, with the risk of falling into the many recesses among the huge masses.

" In the next stage, coral sand has found lodgment among the blocks, and although so scantily supplied as hardly to be detected without close attention, some seeds have taken root, and vines, purslane, and a few shrubs have begun to grow, relieving the scene, by their green leaves, of much of its desolate aspect.

" In the last stage, the island stands six or ten feet out of water. The surface consists of coral sand, more or less discoloured by vegetable or animal decomposition. Scattered among the trees stand, still uncovered, many of the larger blocks of coral, with their usual rough angular features and blackened surface. There is but little depth of coral soil, although the land may appear buried in the richest foliage. In fact the soil is scarcely anything but coral sand. It is seldom discoloured beyond four or five inches, and but little of it to this extent; there is no proper vegetable mould, but only a mixture of darker particles with the white grains of coral sand. It is often rather a coral gravel, and below a foot or two it is usually cemented together into a more or less compact corals and rock."[1]

The animals and plants found on these lagoon islands consist of but few species, and these have been enumerated by Darwin in his "Journal," especially those found upon Keeling Island. "In such a loose, dry, stony soil," he says, "the climate of the intertropical regions alone could produce a vigorous vegetation." "The cocoa-nut palm, at the first glance, seems to compose the whole wood; there are, however, five or six other trees. Besides the trees, the number of plants is exceedingly limited, and consists of insignificant weeds. In my collection, which includes, I believe, nearly the perfect Flora, there are twenty species, without reckoning a moss, lichen, and fungus. To this

[1] Dana, p. 143.

number two trees must be added, one of which was not in flower, and the other I only heard of. As the islands consist entirely of coral, and at one time must have existed as mere water-washed reefs, all their terrestrial productions must have been transported here by the waves of the sea."

"I can hardly explain the reason," he goes on to say, " but there is to my mind much grandeur in the view of the outer shores of these lagoon islands. There is a simplicity in the barrier-like beach, the margin of green bushes, and tall cocoa-nuts, the solid flat of dead coral rock, strewed here and there with great loose fragments, and the line of the furious breakers, all rounding away towards either hand,—the ocean throwing its waters over the broad reef appears an invincible, all-powerful enemy; yet we see it resisted, and even conquered, by means which at first seem most weak and inefficient. It is not that the ocean spares the rock of coral; the great fragments scattered over the reef, and heaped on the beach, whence the tall cocoa-nut springs, plainly bespeak the unrelenting power of the waves. Nor are any periods of repose granted. The long swell caused by the gentle but steady action of the trade wind, always blowing in one direction over a wide area, causes breakers, almost equalling in force those during a gale of wind in the temperate regions, and which never cease to rage. It is impossible to behold these waves without feeling a conviction that an island, though built of the hardest rock, let it be porphyry, granite, or quartz, would ultimately yield, and be demolished by such an irresistible power. Yet these low insignificant coral islets stand, and are victorious; for here another power, as an antagonist, takes part in the contest. The organic forces separate the atoms of carbonate of lime, one by one, from the foaming breakers, and unite them into a symmetrical structure. Let the hurricane tear up its thousand huge fragments, yet what will that tell against the accumulated labour of myriads of architects, at work night and day, month after month? Thus do we see the soft and gelatinous body of a polyp, through the agency of the vital laws, conquering the great mechanical power of the waves of the ocean, which neither the art of man nor the inanimate works of nature could successfully resist."[1]

Allowing a little poetical licence, and remembering that the facts were not so well known in Montgomery's days as in ours, we might add, in the words of the "Pelican Island":—

" Nine times the age of man that coral reef
Had bleach'd beneath the torrid noon, and borne
The thunder of a thousand hurricanes,
Raised by the jealous ocean, to repel
That strange encroachment on his old domain.

[1] Darwin, "Journal of Researches" (1852), p 459.

> His rage was impotent : his wrath fulfill'd
> The counsels of Eternal Providence,
> And 'stablish'd what he strove to overturn :
> For every tempest threw fresh wrecks upon it ;
> Sand from the shoals, exuviæ from the deep,
> Fragments of shells, dead sloughs, sea monsters' bones,
> Whales stranded in the shallows, hideous weeds
> Hurl'd out of darkness by the uprooting surges ;
> These with unutterable relics more,
> Heap'd the rough surface, till the various mass,
> By Nature's chemistry combined and purged,
> Had buried the bare rock in crumbling mould."

At this stage it would be advisable to survey the distribution of coral reefs and atolls, or just as much as will be necessary to comprehend their general localisation. Bearing in mind that the coral itself has been dependent on the living organisms which constructed it, there will at once follow an admission that the temperature of the ocean must be high enough for the organisms to flourish, and here we arrive at the first condition which sets a limit to coral grounds. The reef corals can flourish in the hottest equatorial sea, hence there is no limit in that direction; but it does not appear that they are able to bear a lower winter temperature than 68° Fahr., which seems to be the boundary line on each side of the equator. We need not enter into the particulars why this line is a flexuous one, under the influence of oceanic currents, but so it is, that the line of temperature of the ocean at 68° in winter

CORAL REEFS, AND ISLANDS. 261

is so flexuous in the vicinity of continents, that while the enclosed region is about fifty-six degrees wide in mid-ocean, it is in the Pacific only twenty-five degrees wide on the west coast of America, and forty-five degrees on the Asiatic side, whilst in the Atlantic it is fifteen degrees wide on the African coast, and forty-eight degrees on the coast of America.[1]

Both Darwin and Dana, in their works already alluded to, give abundant details of the geographical distribution of coral reefs, which would scarcely be of much interest to the general reader. We shall be content to allude briefly to some of the most important coral areas, it being understood that we only select a few. In the Pacific Ocean the Paumotas embrace eighty coral islands, and near them the Gambier Islands with extensive reefs. The Society Islands have also extensive coral reefs and barriers, as have also the Samoa or Navigator's Islands. The Tonga Islands for the most part abound in coral reefs. The Fiji group possess reefs

---

[1] The following table is from Prof. Dana's work, and shows the coral boundary lines where they meet continental coasts.

	Pacific.	Atlantic.
East side of Ocean	Lat. 21° N.	Lat. 10° N.
,, ,,	,, 4° S.	,, 5° S.
West side of Ocean	,, 15° N.	,, 26° N.
,, ,,	,, 30° S.	,, 22° S.

of great extent, whilst to the north are numerous islands of coral. The Kingsmill Islands, Marshall Islands, and the Carolines, about eighty in number, are all (excepting three Carolines) atolls. South of the Equator, New Caledonia, and the north-east coast of New Holland is said to be the grandest reef region in the world. Between Australia and New Caledonia the islands are all of coral.

The islands of the Indian ocean are mostly of coral. Such are the Laccadives, Maldives, Keelings, and Chagos Islands. The Seychelles have extensive reefs, and there are fringing reefs on parts of the coast of Madagascar.

In the Atlantic there are few reefs on the American shores, except in the West Indies. The reefs of Florida, Cuba, and the Bahamas are well known. South of the Equator there are reefs at intervals from near Cape St. Roque to the Abrolhos.

It might almost be said that the coral zone girts the world in correspondence with the tropics, whenever the other conditions are favourable to their growth.[1] This reminds us that besides the tempera-

---

[1] I have been assured, by several people, that there are no coral reefs on the West Coast of Africa, or round the islands in the Gulf of Guinea. This, perhaps, may be attributed, in part, to the sediment brought down by the many rivers debouching on that coast, and to the extensive mud banks which line great part of it.—Darwin, "Coral Reefs," p. 62.

ture of the ocean, there are other and important conditions which influence the distribution of corals. One of these is the proximity of the mouths of rivers on account of the sediment which they bring down, and distribute over the sea bottom and the neighbouring coast. No coral reefs can be formed under such conditions. Too steep a shore and too deep water, is another obstruction to the growth of coral. Hereafter it will be shown that corals flourish in comparatively shallow water. Finally, the proximity of volcanic action seems to be a great deterrent to the growth of coral and the development of reefs. For instance, although other conditions are favourable, the island of Hawaii has active volcanoes, and but few traces of coral about it, whereas neighbouring islands, which have long been free from volcanic action, have considerable coral reefs.

The depth at which living corals may be found in the tropics, though varying somewhat in the estimates of different naturalists, is generally admitted to be comparatively small. There was a time when coral reefs were alluded to by voyagers as standing in unfathomable ocean. This might have been only a poetical licence, or, more probably, hazarded without an attempt at sounding. Quoy and Gaymard, the naturalists who explored the Pacific in 1817–1820, were the first to show that the guess was unfounded. According to their observations the limit of distribution downward was from 30 feet to 36 feet. Ehrenberg concluded that in the Red Sea living coral did not occur below 36 feet. Stutchbury, after a visit to the coral groups of the Paumotas and Tahiti, fixed the depth at from 96 feet to 100 feet. Darwin writes that, "Although the limit of depth, at which each particular kind of coral ceases to exist, is far from being accurately known, yet, when we bear in mind the manner in which the clumps of coral gradually became infrequent at about the same depth, and wholly disappeared at a greater depth than 120 feet, on the slope round Keeling atoll, on the leeward side of the Mauritius, and at rather less depth, both without and within the atolls of the Maldive and Chagos Archipelagos, and when we know that the reefs round these islands do not differ from other coral formations, in their form and structure, we may, I , think conclude that in ordinary cases reef-building polypifers do not flourish at greater depths than between 120 feet and 180 feet."[1] Captain Moresby reported to Darwin that he found only decayed coral on Padua Bank, north of the Laccadives which has an average depth of between 150 feet and 210 feet, but that on some other banks in the same group, with only 60 feet to 72 feet of water on them, the coral was living. Lieutenant Wellstead has also

---

[1] Darwin on "Coral Reefs," p. 86.

affirmed that, in the more northern parts of the Red Sea, there are extensive beds of living coral at a depth of 150 feet. Within the lagoons of some of the Marshall atolls, where the water can be but little agitated, there are, according to Kotzebue, living beds of coral in 150 feet. Captain Beechey states that branches of pink and yellow coral were frequently brought up from between 120 feet and 150 feet off the Low atolls, and Lieutenant Stokes, writing from the north-west coast of Australia, says that a strongly-branched coral was procured there from 80 feet. Professor Agassiz observes that, about the Florida reefs, the reef-building corals do not extend below 60 feet. Professor Dana remarks that in the Wilkes's Exploring Expedition the soundings, in the course of the various and extensive surveys, afforded no evidence of growing coral beyond 120 feet. Among the Fiji Islands the extent of coral reef grounds surveyed was many hundreds of square miles. The reefs of the Navigator's Islands were also sounded out, with others of the Society group, and, through all these regions, no evidence was obtained of corals living at a greater depth than 90 feet to 120 feet. He concludes that, "there is hence little room to doubt that twenty fathoms (or 120 feet) may be received as the ordinary limit in depth of reef corals in the tropics."

Not less interesting, or important, is the question

of the growth, or extension of coral, either laterally or vertically, in the formation of reefs, as considered apart from the growth of individual species. On this point all the evidence we can produce is that collected by Darwin, and that of a very limited character. It has been assumed that the vertical growth of coral must be slow, as inferred from a few instances often cited, but which Darwin declares are inconclusive. Ehrenberg has said, for instance, that in the Red Sea, corals only coat other rocks in a layer from one to two feet in thickness, or at most to nine feet, and he did not believe that in any case they would, of their own proper growth, form stratified masses. He alludes to certain large massive corals in the Red Sea, which he imagined of such vast antiquity as to have been cotemporaneous with Pharoah. On this point Darwin admits that there are reefs in the Red Sea which do not appear to have increased in dimensions during half a century, and from comparison of old charts, probably not during the last two hundred years. These, and similar facts, have strongly impressed many with the belief of the extreme slowness of the growth of corals, so that they have even doubted the possibility of islands in the ocean being formed by this agency.

On the contrary, there are facts, from which it may be inferred with certainty, that masses of considerable thickness have been formed by the growth of coral.

There are knolls in the Southern Maldive atolls, some of which, according to Captain Moresby, are less than a hundred yards in diameter, and rise to the surface from a depth of between 250 feet and 300 feet. "Considering their number, form, and position," says Darwin, "it would be preposterous to suppose that they are based on pinnacles of any rock, not of coral formation; or that sediment could have been heaped up into such small and steep isolated cones. As no kind of living coral grows above the height of a few feet, we are compelled to suppose that these knolls have been formed by the successive growth and death of many individuals,—first one being broken off or killed by some accident, and then another, and one set of species being replaced by another set with different habits, as the reef rose nearer the surface, or as other changes supervened. In reefs of the barrier class we may feel sure that masses of great thickness have been formed by the growth of coral; in the case of Vanikoro, judging only from the depth of the moat between the land and the reef the wall of coral rock must be at least 300 feet in vertical thickness."[1] There can be little doubt that Matilda atoll, in the Low Archipelago, has been converted, in the space of thirty-four years, from a "reef of rocks" into a lagoon island, fourteen miles in

---

[1] Darwin, "Structure of Coral Reefs" (1851), p. 72

length, with "one of its sides covered nearly the whole way with high trees." It is urged that, in an old standing reef, the corals are of different kinds at different parts, and all probably adapted to their positions, where they hold their places, like other organic beings, by a struggle with each other and external nature, and hence their growth would generally be slow, except under peculiarly favourable circumstances. "Almost the only natural condition, allowing a quick upward growth of the whole surface of a reef, would be a slow subsidence of the area in which it stood." "If it be asked," says Darwin, " at what rate in years I suppose a reef of coral, favourably circumstanced, would grow up from a given depth, I should answer, that we have no precise evidence on this point, and comparatively little concern with it. We see in innumerable points over wide areas, that the rate has been sufficient, either to bring up the reefs from various depths to the surface, or, as is more probable, to keep them at the surface during progressive subsidences ; and this is a much more important standard of comparison than any cycle of years."

A few facts may be added, and left to carry their own inferences. Dr. Allan experimented in 1830 to 1832 on the east coast of Madagascar. "To ascertain the rise and progress of the coral family, twenty species of coral were taken off the reef, and planted apart on a sandbank, *three feet deep at low-water.* Each portion weighed ten pounds, and was kept in its place by stakes. Similar quantities were placed in a clump, and secured as the rest. This was done in December, 1830. In July following each detached mass was nearly level with the sea, at low-water, quite immovable, and several feet long, stretching as the parent reef, with the coast current from north to south."

Lieutenant Wellstead communicated the fact that " in the Persian Gulf a ship had her copper bottom encrusted, in the course of twenty months, with a layer of coral *two feet* in thickness, which it required great force to remove when the vessel was docked."

The inhabitants at Keeling Island made with crowbars a considerable channel through the reefs, in which a schooner was floated out. In less than ten years this channel was again almost choked up with living coral, so that fresh cuttings would be absolutely necessary before another vessel could pass through.

If these instances can be accepted as types, or as approximations towards the determination of the growth of coral, under ordinary circumstances and conditions, then it can scarcely be maintained that the growth of coral is always slow.

The most popular of the older theories of the formation of coral-atolls, was that which regarded them as crowning the submerged summits of extinct volcanoes. The lagoon in the centre corresponded to the crater of the volcano, and the belt of land encircling the lagoon was the rim of the crater. This theory was rendered the more probable from the volcanic nature of the region in which the best known atolls had been found. This was the only theory which held any position in the world when Darwin announced his hypothesis. The objections to the volcanic theory have been summarised by Dana:—

(1) The volcanic cones must either have been subaërial, and then have afterward sunk beneath the waters, or else they were submarine from the first. In the former case the crater would have been destroyed, with rare exceptions, during the subsidence; and in the latter there is reason to believe that a distinct crater would seldom, if ever, be formed.

(2) The hypothesis moreover requires that the ocean's bed should have been thickly planted with craters,—seventy in a single archipelago,—and that they should have been of nearly the same elevation, for, if more than twenty fathoms below the surface, corals could not grow upon them. But no records warrant the supposition that such a volcanic area ever existed. The volcanoes of the Andes differ from 1,000 feet to 10,000 feet in altitude, and scarcely two cones throughout the world are as nearly of the same height as here supposed. Mount Loa and

Mount Koa, of Hawaii, present a remarkable instance of approximation, as they differ but 200 feet; but the two sides of the crater of Mount Loa differ 314 feet in height. Mount Koa, though of volcanic character, has no large crater at top. Hualalai, the third mountain of Hawaii, is 4,000 feet lower than Mount Loa. The volcanic summit of East Maui is 10,000 feet high, and contains a large crater, but the wall of the crater, on one side, is 700 feet lower than the highest point of the mountain, and the bottom of the crater is 2,000 feet below the rim of the crater. Similar facts are presented by all volcanic regions.

(3) It further requires that there should be craters over fifty miles in diameter, and that twenty and thirty miles should be a common size. Facts give no support to such an assumption.

(4) It supposes that the high islands of the Pacific, in the vicinity of the coral islands, abound in craters; while on the contrary, there are none, so far as is known, in the Marquesas, Gambier, or Society Group, the three which lie nearest to the Paumotas. Even this supposition fails, therefore, of giving plausibility to the crater hypothesis.[1] It is no exaggeration to claim that this ancient theory, having become utterly untenable, has been everywhere abandoned.

---

[1] Dana, "Corals and Coral Islands," p. 219.

It remains briefly to sketch the hypothesis advanced by Darwin, sanctioned by Humboldt, supported by Dana, and generally adopted. This theory is based upon the fundamental fact of the gradual subsidence, or elevation, of islands in the Pacific, that is to say, on changes of level in the Pacific Ocean. Professor Dana devotes an entire chapter in his work to the substantiation, and elucidation, of this fundamental position. Those interested in ascertaining the data for this position must consult that volume, as it would occupy too much space to recapitulate here. Further, that the extent of subsidence has been very considerable. That, in connexion with the origin of coral islands, and barrier reefs, in the Pacific, it must have amounted to several thousands of feet, perhaps fully ten thousand. And it is further urged that this change of level is not greater than the elevation which the Rocky Mountains, Andes, Alps, and Himalayas, have each experienced since the close of the Cretaceous era, or the early Tertiary; and perhaps it does not exceed the upward bulging in the Glacial era of part of northern North America.

Humboldt gives the briefest, and most succinct, summary of this hypothesis, when he writes "According to Darwin the following is the process of formation. An island mountain closely encircled by a coral reef subsides, while the fringing reef, that had sunk with it, is constantly recovering its level, owing to the tendency of the coral animals to regain the surface, by renewed perpendicular structures; these constitute, first a reef encircling the island at a distance, and subsequently, when the inclosed island has wholly subsided, an atoll. According to this view, which regards islands as the most prominent parts, or the culminating points of the submarine land, the relative position of the coral islands would disclose to us, what we could scarcely hope to discover by the sounding line, viz., the former configuration and articulation of the land."[1]

Admirable as this short outline of the theory may be it is scarcely sufficient, and needs to be supplemented by Darwin's own account of his theory, as set forth in his "Journal," and still further expanded in his larger work.

"We have seen that we are driven to believe in the subsidence of those vast areas, interspersed with low islands, of which not one rises above the height to which the wind and waves can throw up matter, and yet are constructed by animals requiring a foundation, and that foundation to lie at no great depth. Let us then take an island, surrounded by fringing reefs, which offer no difficulty in their structure, and let this island with its reef, represented by the unbroken line in the woodcut (fig. 54), slowly

---

[1] Humboldt's "Views of Nature," Illustrations, p. 262. London, 1850.

subside. Now, as the island sinks down, either a few feet at a time, or quite insensibly, we may safely infer, from what is known of the conditions favourable to the growth of coral, that the living masses, bathed by the surf on the margin of the reef, will soon regain the surface. The water, however, will encroach little by little on the shore, the island becoming lower and smaller, and the space between the inner edge of the reef and the beach proportion-

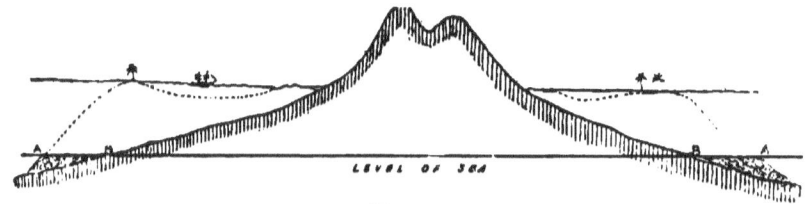

FIG. 54.

A. A. Outer edges of fringing reef.
B. B. Shores of fringed island.
Upper horizontal line showing sea level after a period of subsidence.

ally broader. A section of the reef and island in this state, after a subsidence of several hundred feet, is given by the dotted lines. Coral islets are supposed to have formed on the reef, and a ship is anchored in the lagoon channel. This channel will be more or less deep, according to the rate of subsidence, to the amount of sediment accumulated in it, and to the growth of the delicately-branched corals which can live there. The section, in this state, resembles in every respect one drawn through an encircled island, in fact, it is a real section, through Bolabola in the Pacific. We can now at once see why encircling barrier reefs stand so far from the shores which they front. We can also perceive, that a line drawn perpendicularly down from the outer edge of the new reef to the foundation of solid rock beneath the old fringing reef, will exceed, by as many feet as there have been feet of subsidence, that small limit of depth at which the effective corals can live—the little architects having

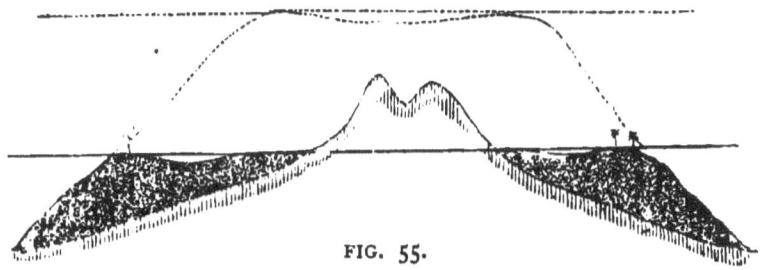

FIG. 55.
Lower horizontal line showing sea level with islets on it.
Upper horizontal line, the sea level after further subsidence,
with the reef converted into an atoll.

built up their great wall-like mass, as the whole sank down, upon a basis formed of other corals, and their consolidated fragments. Thus the difficulty on this head, which appeared so great, disappears.

" Let us take our now encircling barrier reef, of which the section is now represented by unbroken lines, and which, as I have said, is a real section through Bolabola, and let it go on subsiding (fig. 55). As the barrier reef slowly sinks down, the corals will go on vigorously growing upwards ; but, as the island sinks, the water will gain inch by inch on the shore, the separate mountains first forming separate islands within one great reef, and finally, the last and highest pinnacle disappearing. The instant this takes place a perfect atoll is formed ; I have said, remove the high land from within an encircling barrier-reef, and an atoll is left, and the land has been removed. We can now perceive how it comes that atolls, having sprung from encircling barrier-reefs, resemble them in general size, form, in the manner in which they are grouped together, and in their arrangement in single or double lines ; for they may be called rude outline charts of the sunken islands over which they stand. We can further see how it arises that the atolls, in the Pacific and Indian oceans, extend in lines parallel to the generally prevailing strike of the high islands, and great coast-lines of those oceans. I venture, therefore, to affirm, that on the theory of the upward growth of the corals, during the sinking of the land, all the leading features in those wonderful structures, the lagoon islands, or atolls, which have so long excited the attention of voyagers, as well as in the no less wonderful barrier-reefs, whether encircling small islands, or stretching for hundreds of miles along the shores of a continent, are simply explained."[1]

---

[1] Darwin, "Journal of Researches," p. 473 (1852).

The Duke of Argyll, when controverting the Darwinian theory of subsidence, admits that "the theory of the young naturalist was hailed with acclamation. It was a magnificent generalisation. It was soon almost universally accepted with admiration and delight. It passed into all popular treatises, and ever since, for the space of nearly half a century, it has maintained its unquestioned place, as one of the great triumphs of reasoning and research. Although its illustrious author has since eclipsed this earliest performance, by theories and generalisations still more attractive, and much further reaching, I have heard eminent men declare, that, if he had done nothing else, his solution of the great problem of the coral islands of the Pacific would have sufficed to place him on the unsubmergeable peaks of science, crowned with an immortal name."[1]

Following immediately upon this concession comes the "great lesson" which the noble Duke deems it to be his mission to teach. "After an interval of more than five-and-thirty years, the voyage of the *Beagle* has been followed by the voyage of the *Challenger*, furnished with all the newest appliances of science, and manned by a scientific staff, more than competent to turn them to the best

---

[1] "A Great Lesson," by the Duke of Argyll, in *The Nineteenth Century*, No. 127. Sept. 1887, p. 300.

account. And what is one of the many results which have been added to our knowledge of nature, to our estimate of the true character and history of the globe we live on? It is that Darwin's theory is a dream. It is not only unsound, but it is, in many respects, directly the reverse of truth. With all his conscientiousness, with all his caution, with all his powers of observation, Darwin in this matter fell into errors as profound as the abysses of the Pacific. All the acclamations, with which it was received, were as the shouts of an ignorant mob. It is well to know that the plebiscites of science may be as dangerous, and as hollow, as those of politics. The overthrow of Darwin's speculation is only beginning to be known. It has been whispered for some time. The cherished dogma has been dropping very slowly out of sight. Can it be possible that Darwin was wrong? Must we, indeed, give up all that we have been accepting and teaching for more than a generation? Reluctantly, almost sulkily, and with a grudging silence, as far as public discussion is concerned, the ugly possibility has been contemplated, as too disagreeable to be much talked about. The evidence, old and new, has been weighed, and weighed again, and the obviously inclining balance has been looked at askance many times. But, despite all averted looks, I apprehend that it has settled to its place for ever, and Darwin's theory of the coral islands must be relegated to the category of those many hypotheses which have, indeed, helped science for a time, by promoting and provoking further investigation, but which, in themselves, have now finally 'kicked the beam.'

"But this great lesson will be poorly learnt unless we read and study it in detail. What was the flaw in Darwin's reasoning, apparently so close and cogent? Was it in the facts, or was it in the influences? His facts in the main were right; only it has been found that they fitted into another explanation better than into his. It was true that the corals could only grow in a shallow sea, not deeper than from twenty to thirty fathoms. It was true that they needed some foundation provided for them, at the required depth. It was true that this foundation must be in the pure and open sea, with its limpid water, its free currents, and its dashing waves. It was true that they could not flourish, or live in lagoons, or in channels, however wide, if they were secluded and protected from oceanic waves. One error, apparently a small one, crept into Darwin's array of facts. The basis, or foundation, on which corals can grow, if it satisfied other conditions, need not be solid rock. It might be deep-sea deposits, if these were raised or elevated near enough the surface. Darwin did not know this, for it is one of his assumptions that coral 'cannot adhere to a loose bottom.' The *Challenger* observations show that thousands of deep-sea corals, and of other lime-secreting animals, flourish on deep-sea deposits, at depths much greater than those at which true reef-building species are found. The dead remains of these deeper-living animals, as well as the dead shells of pelagic species, that fall from the surface waters, build up submarine elevations towards the sea-level. Again, the reef-building coral will grow upon its own debris,—rising, as men, morally and spiritually, are said by the poet to do,—

"'On stepping-stones of their dead selves to higher things.'

This small error told for much ; for if coral could grow on deep-sea deposits when lifted up, and if it could also grow seaward, when once established, upon its own dead and sunken masses, then submarine elevations, and not submarine subsidences, might be the true explanation of all the facts. But what of the lagoons, and the immense areas of sea behind the fringing reefs? How could these be accounted for? It was these which first impressed Darwin with the idea of subsidence. They looked as if the land had sunk behind the reef, leaving a space into which the sea had entered, but in which no fresh reefs could grow. And here we learn the important lesson, that an hypothesis may adequately account for actual facts, and yet, nevertheless, may not be true. A given agency may be competent to produce some given effect, and yet that effect may not be due to it, but to some other. Subsidence would, or might, account for the lagoons, and for the protected seas, and yet it may not be subsidence which has actually produced them.

" Darwin's theory took into full account two of the great forces which prevail in nature, but it took no account of another, which is comparatively inconspicuous in its operations, and yet is not less powerful than the vital energies, and the mechanical energies, which move and build up material. Darwin had thought much and deeply on both of these. He called on both to solve his problem. To the vital energy of the coral animals he rightly ascribed the power of separating the lime from the sea-water, and of laying it down again in the marvellous structures of their calcareous homes. In an eloquent and powerful passage, he describes the wonderful results, which this energy achieves, in constructing breakwaters, which repel and resist the ocean, along thousands of miles of coast. On the subterranean forces which raise, and depress, the earth's crust he dwelt,—at least enough. But he did not know, because the science of his day had not then fully grasped, the great work performed by the mysterious power of chemical affinity, acting through the cognate conditions of aqueous solution. Just as it did not occur to him that a coral reef might advance steadily seaward, by building ever fresh foundations on its own fragments, when broken and submerged, or that the vigorous growth of the reefs, to windward, was due to the more abundant supply of food, brought to the reef-building animals from that direction, by oceanic currents, so did it never occur to him that it might melt away to the rear, like salt or sugar, as the vital energy of the coral animals failed, in the sheltered and comparatively stagnant water. It was that vital energy alone, which not only built up the living tubes and cells, but which filled them with living organic matter, capable of resisting the chemical affinities of the inorganic world. But when that energy became feeble, and when at last it ceased, the once powerful structure descended again to that lower level of the inorganic, and subject to all its laws. Then, what the ocean could not do, by the violence of its waves, it was all-potent to do by the corroding and dissolving power of its calmer lagoons. Ever eating, corroding, and dissolving, the back waters of the original fringing reef—the mere pools and channels left by the outrageous sea, as it dashed upon the shore—were ceaselessly at work, aided by the high temperature of exposure to blazing suns, and by the gases evolved from decaying organisms. Thus the enlarging area of these pools and channels spread out into wide lagoons, and into still wider protected seas. They needed no theory of subsidence to account for their origin or for their growth. They would present the same appearance in a slowly-rising, a stationary, or a slowly-sinking area. Their outside boundary was ever marching further outward, on submarine shoals and banks, and ever, as it advanced in that direction, its rear ranks were melted and dissolved away. Their inner boundary—the shores of some island, or of some continent—might be steady and unmoved, or it might be even rather rising instead of sinking. Still, unless this rising were such as to overtake the advancing reef, the lagoon would grow, and if the shores were steady, it would widen as fast as the face of the coral barrier could advance. Perhaps, even if such a wonderful process had ever occurred to Darwin,—even if he had grasped this extraordinary example of the 'give and take' of nature,—of the balance of opposing forces, and agencies, which is of the very essence of its system, he would have been startled by the vast magnitude of the operations which such an explanation demanded. In its incipient stages this process is not only easily conceivable, but it may be seen in a thousand places, and in a thousand stages of advancement. There are islands, without number, in which the fringing reef is still attached to the shore, but in which it is being 'pitted,' holed, and worn into numberless pools on the inner surfaces, where the coral is in large patches dead or dying, and where its less soluble ingredients are being deposited in the form of coral sand. There are thousands of other cases where the lagoon interval, between the front of the reef and the shores, has been so far widened that it is taking the form of a barrier, as distinguished from a fringing reef, and where the lagoon can be navigated by small boats. But when we come to the larger atolls, and the great seas included between a barrier reef and its related shores, the mind may well be staggered by the enormous quantity of matter, which it is suggested has been dissolved, removed, and washed away. The breadth of the sheltered seas, between barrier reefs and the shore, is measured in some cases not by yards, nor hundreds of yards, not by miles, but by tens of miles, and this breadth is carried on in linear directions, not for hundreds of miles, but for thousands. And yet there is one familiar idea in geology which might have helped Darwin, as it is much needed to help us, even now, to conceive it. It is the old doctrine of the science, long ago formulated by Hutton, that the work of erosion, and of denudation, must be equal to the work of deposition. Rocks have been formed out of the ruins of older rocks, and those older rocks must have been worn down, and carried off, to an equivalent amount. So it is here, with another kind of erosion, and another kind of deposition. The coral building animals can only get their materials from the sea, and the sea can only get its materials by dissolving it from calcareous rocks of some kind. The dead corals are amongst its greatest quarries. The inconceivable, and immeasurable quantities, which have been dissolved out of the lagoons, and sheltered seas of the Pacific, and of the Indian Ocean, are not greater than the immeasurable quantities which are again used up, in the vast new reefs of growing coral, and in the calcareous covering of an inconceivable number of other marine animals."

Such a conception the noble Duke thinks may "not be so imposing as that of a whole continent gradually subsiding, of its long coasts marked by barrier reefs, of its various hills, and irregularities of surface, marked by islands of corresponding size, and finally, of the atolls indicating where its highest peaks finally disappeared beneath the sea. But, on the other hand," he considers, " the new explanation more like the analogies of nature,—more closely correlated with the wealth of her resources, with those curious reciprocities of service, which all her agencies render to each other, and which indicate so strongly the ultimate unity of her designs."

We have considered it advisable to give in full, and in his own language, the impeachment of Darwin's theory, and elucidation of his own, which the noble author has advanced. Albeit, it is confessed, that for the new theory he is indebted to Mr. John Murray, one of the naturalists of the *Challenger* expedition, as will be apparent from the subjoined extract, which sets forth briefly the views entertained by him on the subject, and at first communicated to the Royal Society of Edinburgh, in 1880.

"According to Mr. Murray, the observations of the reefs at Tahiti support the view that the reefs have been built from the shore seawards, and that the lagoons have been, and are still being, formed by the removal of the inner, and dead portions of the coral reef, by the solvent action of sea water. The islands in the harbour, and lagoons, are regarded as portions of the reef which have been left standing, but will ultimately be removed, and, in confirmation of this, it is pointed out, that on the inner part of the reef there are large and massive specimens of the coral which are now dead, but which probably flourished at the time when the outer edge of the reef was at the position in which they are now found. The steep slope which is found on the outer edge of the reef, between the depths of 35 and 200 fathoms, is believed to be formed by huge masses and heads of coral, which have been torn away from the ledge, between the edge of the reef and 35 fathoms, during storms, or by overhanging masses which have fallen by their own weight. In this way a talus has been formed on which the corals, living down to 35 fathoms, have found a foundation on which to build further seawards, for this seaward slope is the great growing surface of the reef. The food supply for the masses of living coral, on the outer slope of the reef, is brought by the oceanic currents sweeping past the islands, a fact in relation with the more vigorous growth of the reef on the windward sides. It is maintained by Mr. Murray that the whole of the phenomena of the Tahiti reefs may be fully explained, by reference to the processes at present in action, and without calling in the aid of subsidence, as is done by Darwin and Dana, and, it is argued further, that the form of atoll, and barrier reefs generally, can be explained on the same principles."[1]

Having endeavoured, without prejudice, to set forth the theory propounded by Darwin, and supported by Dana, and others, as well as the countertheory, advanced by Murray, and advocated by the Duke of Argyll, it will be expected of us to impart the result of our own convictions in a few concluding lines. Those who followed the controversy in the pages of *Nature*, and elsewhere, need not to be reminded that the verdict of just those scientific men who were most capable of judging, was *not* heard, in condemnation of the hypothesis advanced

---

[1] "Narrative of the Voyage of the *Challenger*," vol. i. p. 781 (1885); also "On the Structure and Origin of Coral Reefs and Islands," by John Murray, in *Proc. Roy. Soc. Edin.*, x. p. 505 (1880).

by Darwin, and in support of that proposed by Murray. It was evident that the new theory failed to satisfy the majority, it being insufficient to account for all the phenomena, as the old theory had done. Without actually calling in question the facts, the inferences were not held to be warranted, and there were still remaining certain phenomena which the new theory could not account for.

This reminds us of some observations made by Professor Huxley on the subject of hypotheses, which should always be borne in mind. "Wherever," he says, "there are complex masses of phenomena to be inquired into, whether they be phenomena of the affairs of daily life, or whether they belong to the more abstruse and difficult problems laid before the philosopher, our course of proceeding in unravelling that complex chain of phenomena, with a view to get at its cause, is always the same; in all cases we must invent an hypothesis; we must place before ourselves some more or less likely supposition respecting that cause; and then, having assumed an hypothesis, having supposed a cause for the phenomena in question, we must endeavour, on the one hand, to demonstrate our hypothesis, or, on the other, to upset and reject it altogether, by testing it in three ways. We must, in the first place, be prepared to prove that the supposed causes of the phenomena exist in nature, that they are what the logicians call *vera causæ,*—true causes ; in the next place, we should be prepared to show that the assumed causes of the phenomena are competent to produce such phenomena as those which we wish to explain by them ; and in the last place, we ought to be able to show that no other known causes are competent to produce these phenomena. If we can succeed in satisfying these three conditions we shall have demonstrated our hypothesis ; or rather, I ought to say, we shall have proved it as far as certainty is possible for us ; for, after all, there is no one of our surest convictions which may not be upset, or, at any rate, modified by a further accession of knowledge."[1] And again :— " Every hypothesis is bound to explain, or at any rate, not be inconsistent with, the whole of the facts which it professes to account for ; and if there is a single one of these facts which can be shown to be inconsistent with the hypothesis, the hypothesis falls to the ground—it is worth nothing. One fact with which it is positively inconsistent is worth as much, and as powerful in negativing the hypothesis, as five hundred." Tested by these obligations, it hardly seems clear that the new hypothesis is competent to supersede the old one. At least it is by no means so

---

[1] "On our Knowledge of the Causes of the Phenomena of Organic Nature." By Professor J. Huxley. London, 1863, p. 135.

perfect a hypothesis, it does not reconcile so many facts, or account for so many phenomena, and, provided that the new facts are established, it is by no means certain that they are inconsistent with the original hypothesis of subsidence. If the new hypothesis is really so much more conclusive, so much more feasible, so much more consistent than the old one, how does it happen that it has failed, and still continues to fail, in securing the support of the thinking and reasoning public, especially that portion which are specially interested, and most competent to judge? Surely, if the theory had been so clear, so convincing, so evident, as some have contended, a period of eight years is sufficient to have established it in public opinion. Whatever its ultimate destiny may be, we recognise no valid reason, at present, for rejecting the hypothesis of Darwin and Dana, in favour of that propounded, or supported, by Murray and Argyll.

CHAPTER IX.

SEA-MAT MAKERS.

ONE of the most common experiences of a sea-side collector is to pick up a dead frond of the larger sea-weeds, or a tuft of the smaller ones, wholly, or only partially, invested with an encrusting patch, of a dirty-white colour, evidently of quite a different character from the sea-weed on which it has become established ; and, if examined by the pocket-lens, will be found to consist of a great number of horny cells, each with a small opening or mouth, and, in some species, furnished with long rigid bristles, so as appear even to the naked eye to be hairy. These were once the homes of minute marine Polyzoa, kindred to the sea-mats, the latter differing from the former in not encrusting other substances, but forming a kind of frond, or tuft of fronds, of their own. The Polyzoa, or as some persons persist in calling them, the Bryozoa, are for the most part marine, although there is a small group, well known to pond-hunters, which inhabit fresh water ; in all of them the animals are small, but the colonies, or aggregations of individuals, are often of considerable size. All of them appear to be gregarious, flourishing in companies, and the number of species must be very large, for they have a wide, and almost unlimited, distribution. Along our own shores it would be difficult to say where we should find no trace of them. Some of the encrusting species are found flourishing on almost every dead shell, and often whilst the mollusk is still living; others flourish on stones and rocks which are submerged, and a large number attach themselves to sea-weeds. They will be found encrusting wood, stones, crabs, lobsters, or attached to zoophytes, or even other species of Polyzoa, and indeed seem to be almost ubiquitous in the sea. Their general structure it must be our first endeavour to attempt to elucidate.

There is a superficial resemblance in some of the species to the common Hydroid zoophytes, but, examined more closely, the animal will be found to be of the molluscan type, and not that of the Hydroids.

Elsewhere we have given a summary of the general type of Polyzoon,[1] after Professor Allman, to the following effect:—" Let us imagine an alimentary canal, consisting of œsophagus, stomach and intestine,

---

[1] " Natural History Rambles : Ponds and Ditches," p. 132, 1880.

to be furnished at its origin with long ciliated tentacles, and to have a single nervous ganglion situated on one side of the œsophagus (fig. 56). Let us now suppose this canal to be bent back upon itself, towards the side of the ganglion, so as to approximate the termination to the origin. Further, let us imagine the digestive tube, thus constituted, to be suspended in a fluid, contained in a membraneous sac, with two openings, one for the mouth and the other for the vent, the tentacles alone being external to the sac. Let us still further suppose the alimentary tube, by means of a system of muscles, to admit of being retracted or protruded, according to the will of the animal, the retraction being accompanied by an invagination (or folding inwards) of the sac, so as partially, or entirely, to include the oral tentacles within it, and if to these characters we add the presence of true sexual organs, occupying some portion of the interior of the sac, and the negative character of the absence of all vestige of a heart, we shall have, perhaps, as correct an idea as can be conveyed of the essential structure of a Polyzoon, in its simplest and most generalised condition."

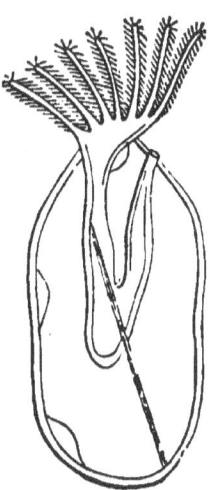

FIG. 56.
TYPICAL POLYZOON.

The cell, or polype home (called scientifically *Zoœcium*), has a double cell-wall, enclosing the initial polyzoon. The outer coat is a chitinous, or horny, membrane, thickened and strengthened, in most cases, by deposits of lime or flint, then forming a solid wall, externally often ornamented or embossed, but occasionally it retains its membraneous character. The colony consists of a number of these cells, attached together in a variety of ways, but produced by consecutive budding, or gemmation, from the original cell. Communication is maintained between the several cells of the colony, through thin perforated plates in the outer wall, called "communication-plates." Through the minute perforations in these plates thread-like connectors pass from animal to animal, and thus all the polypes in a colony are linked together. The function of this outer wall, then, is the protection of the enclosed polype.

The inner wall of the cell is a soft living membrane, which lines the more or less solid outer wall, and is supposed to be intimately associated with all budding processes which subsequently take place. It will be sufficient here to regard it as the soft lining of the walls of the house which contains the living Polyzoon.

The animal is crowned with a wreath of tentacles, which are seated on a circular disc, surmounting the body, with the mouth in the centre (fig. 57). These

SEA-MAT MAKERS.

tentacles when expanded form a bell-shaped wreath, and are slender hollow filaments, with a line of movable cilia down two opposite sides, which, by their incessant motion, cause an inward current of the water, and sweep the floating particles of food into the mouth at the centre. In number the tentacles vary from eight to eighty, and are very energetic in their movements. They are probably the sole organs of touch which the animals possess.

The sheath is a membraneous extension, from the upper extremity of the polype cell, to the base of the crown of tentacles, when expanded, and when the tentacles are drawn in the sheath is drawn in with them. "The invagination of the sheath is due to its attachment, at its upper extremity, round the base of the crown of tentacles; as the latter descends, in obedience to the summons of the retractor muscles, it is, of course, drawn down with it, and reversed as the finger of a glove might be under similar circumstances, forming a protective case around it, which sometimes exceeds it in length. By this arrangement the cell (*Zoœcium*) is completely closed; there

FIG. 57.
BRANCHED TUBULAR POLYZOA
(*Fredericella*).

is no real opening through which the polype passes. It is the upward movement of the tentacular crown which carries with it, and everts, the flexible sheath, and so permits the imprisoned zooid a certain amount of communication with the outer world; but the cavity of the cell itself is sealed. The movements of the polype, in the acts of expansion and retraction, are limited to the eversion and inversion of the sheath; and only the crown is brought into immediate contact with the surrounding water"[1] (fig. 58).

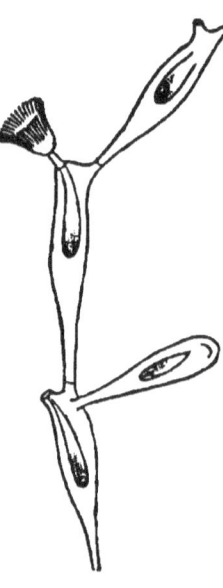

FIG. 58.
CREEPING POLYZOA
(*Paludicella*).

There is nothing which calls for special notice, or description, here of the stomach and its digestion, of the nervous or muscular system, and other minute anatomical details, which would be essential for the student, but rather let us seek to learn something of the "brown bodies" as they are called, which have been the cause of considerable controversy. Let us take, for example, one of the erect growing

---

[1] "History of the British Marine Polyzoa," by Thomas Hincks, London, 1880, vol. i. p. 17.

Polyzoa, with somewhat the form of a zoophyte. The terminal and outside cells will be those of young polypes, and under, or adjoining these, mature and vigorous individuals, below these will be found cells for the most part destitute of tenants. In the majority of these dead cells will be found a dark spherical body, of brownish substance, enclosed within a membraneous cyst, or bladder, and this has been called the "brown body"; two or more may be found in the same cell. What are these "brown bodies," and what are their functions? "They have been regarded as the remains of dead polypes, as ova, as ovaria, as statoblasts, as a secretion from the endocyst, as a store of nutriment for the young polypes, as a reproductive body formed out of the stomach of the decaying polype."[1] Observers are generally agreed in that it is derived from the substance of the polype. Certain writers have contended that it is nothing more than the remains of the old polype, only a mass of inert material, waiting to be ejected from the cell. Whatever it may prove to be, it has undoubtedly been derived from the polype, which inhabited the cell, and is the result of its decline. Another class of authors contend that the "brown body" is capable of originating a new

---

[1] See "Contributions to the History of the Polyzoa," by T. Hincks, in *Quart. Journ. Micr. Science*, vol. xiii. (n.s.) p. 24.

polype. This view seems to have the preponderance of evidence in its favour. One author claims to have traced the formation of a bud on the "brown body," and, as it seemed, out of its very substance. Mr. Hincks says that, "repeatedly I have seen gemmation taking place, in the closest proximity to the surface of the "brown body," and the bud, as I was fully persuaded at the time, was continuous with it, and a growth out of it." "Another point my observations seemed to me to have established,—that the polypes developed from the (so-called) germ capsule, differ in appearance during their early stages from those which are found in the young marginal cells of the colony, and from other buds which occur in the adult polype cell (*Zoæcia*). The latter are destitute of the reddish-brown colour, imparted to the stomach wall at a later period by the biliary glands, whilst the former, being a growth out of the brown body, possess it from the first." To these views it has been objected that the bud is not developed from the "brown body," but from the living substance which surrounds it; and that at a certain stage the bud approaches the "brown body," extends its substance over it, and finally lodges it in its stomach, the walls of which completely close around it. Thus engulfed, it serves as a store of nutriment for the growing structure. As a summary of what appears to have been ascertained, Mr. Hincks concludes,

"the following points may be taken as established :—
(1.) The brown body is universally derived from the substance of the histolyzed polype. (2.) It is always attached (when occupying its original position) to the funiculus, and more or less invested by the funicular plexus. (3.) Buds for the production of a new polype are very commonly developed in the closest proximity to it, and on its surface. (4.) They also originate in various positions, and at greater or less distances from the brown body. (5.) In some species the latter is enveloped by the neighbouring bud, and passes into the digestive canal, being ultimately expelled through the intestine, either entire, or after having undergone dissolution in the stomach. (6.) There may be several brown bodies in a cell; and in some cases they lie loose in the perivisceral cavity near the bottom of it."[1] Hence the final issue has apparently still to be decided, and the contradictory facts reconciled, that the same body may, in some cases, be an important reproductive factor, and at another ejected as mere rubbish.

Singular appendages, called "bird's-head processes," are found attached to the sides of the cells in some species of the Sea-mats. Their functions are very imperfectly known, but their appearance attracted the attention of Darwin when on his voyage,

---

[1] Hincks, "British Polyzoa," vol. i. p. 63.

and his remarks are still applicable He says,—" The organ, in the greater number of cases, very closely resembles the head of a vulture (fig. 59); but the lower mandible can be opened much wider, so as to form even a straight line with the upper. The head itself possesses considerable powers of movement, by means of a short neck. In one zoophyte the head itself was fixed, but the lower jaw free; in another it was replaced by a triangular hood, with a beautifully-fitted trap-door, which evidently answered to the lower mandible. A species of stony *Eschara* had a structure somewhat similar. In the greater number of species, each cell was provided with one head, but in others each had two.

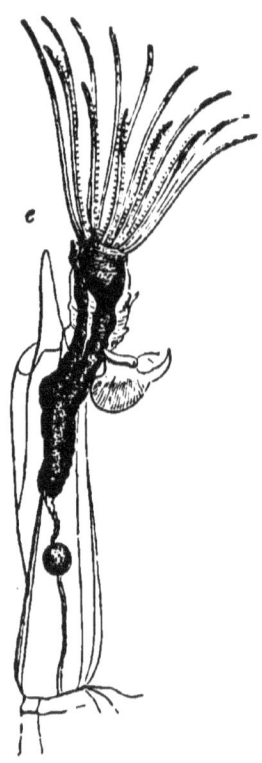

FIG. 59.
CELLULARIA AVICULARIA.

" The young cells at the end of the branches necessarily contained quite immature polypes, yet the vulture-heads attached to them, though small, were in every respect perfect. When the polyp was removed by a needle, from any of the cells, the organs did not appear in the least affected. When one of the latter was cut off from a cell, the lower mandible retained its power of opening and closing. Perhaps the most singular part of their structure is, that when there were more rows of cells than two, the central cells were furnished with these appendages, of only one-fourth the size of the lateral ones. Their movements varied according to the species; in some I never saw the least motion; while others, with the lower mandible generally wide open, oscillated backwards and forwards, at the rate of about five seconds each turn; others moved rapidly and by starts. When touched with a needle the beak generally seized the point so firmly, that the whole branch might be shaken.

"These bodies have no relation whatever with the production of the gemmules. I could not trace any connexion between them and the polyp. From their formation being completed before that of the latter; from the independence of their movements; from the difference of their size in different parts of the branch :—I have little doubt that in their functions they are related rather to the axis than to any of the polyps."

He then proceeds to notice another kind of structure which is analogous to the vulture-heads, and which some naturalists consider to be only a form or modification of the "bird's-head processes." He proceeds :—" A small and elegant species is furnished, at the corner of each cell, with a long and slightly curved bristle, which is fixed at the lower end by a joint. It terminates in the finest point, and has its outer, or convex, side serrated with delicate teeth or notches. Having placed a small piece of a branch under the microscope, I was exceedingly surprised to see it suddenly start from the field of vision, by the movement of these bristles, which acted as oars. Irritation generally produced this motion, but not always. When the coral was laid flat on that side from which the toothed bristles projected, they were necessarily all pressed together and entangled. This scarcely ever failed to excite a considerable movement among them, and evidently with the object of freeing themselves. In a small piece which was taken out of water, and placed on blotting-paper, the movements of these organs were clearly visible for a few seconds by the naked eye.

"In the case of the vulture-heads, as well as in that of the bristles, all that were on one side of the branch, moved sometimes co-instantaneously, sometimes in regular order, one after the other; at other times the organs on both sides of the branch moved together; but generally all were independent of each other, and entirely so of the polyps. If the bristles were excited to move, by irritation, in any one branch, generally the whole zoophyte was affected. In the instance where the branch started from the simultaneous movement of these appendages, we see as perfect a transmission of will as in a single animal."[1]

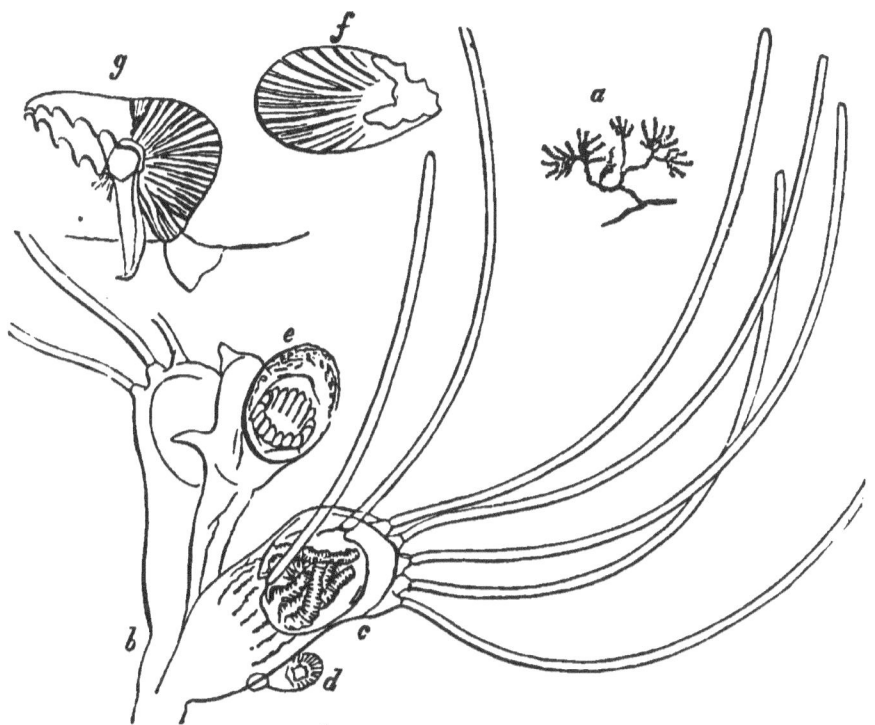

FIG. 60.—CELLULARIA CILIATA.
(*a*) Natural size.   (*b*) Magnified.   (*g*) Bird's-head process.

The "bird's-head processes" are not present in all species, nor on all specimens of the same species, and when they are present it is only in connexion

---

[1] Darwin, "Voyages of *Adventure* and *Beagle*," vol. iii. p. 259.

with some of the cells (fig. 60). They are classified by Krohn under three kinds. Those which constitute the proper "bird's-head process," those which resemble pincers, and those which partake of the form of bristles or hairs. Van Beneden traced their growth and development, without obtaining any clue to their use or purpose. Dr. Reid describes the bristles in a common British species thus :—" At the anterior part of the outer side of each cell (in *Cellularia scruposa*) and immediately in front of the toothlike process there attached, are two pretty long spines, and a rounded process, which tapers slightly from its fixed to its free extremity. This rounded process is open at the top, and is hollow in dead specimens, but when alive it is full of a contractile substance. In this contractile substance the end of a hair-like curved filament, about the length of the cell, is immersed. This hair-like filament is moved about, by the contractile substance attached to it, generally in jerks, after intervals of repose, and in its movements sweeps the anterior and posterior surfaces of the cell to which it is fixed. These movements continue for a considerable time after the animal inhabiting the cell has been dead. A hollow rounded process, with a hair-like curved and moveable filament projecting from it, is also fixed upon the corresponding part of each cell. These moveable hair-like filaments are analogous to the moveable bird's-head process attached to each of the cells of the sea-mat (*Flustra avicularis*)."[1]

Mr. Hincks has traced the modification of the *avicularium* and *vibraculum*, as these processes are called, in a large number of species, and came to the conclusion that they are, morphologically, to be regarded as metamorphosed polype-cells (*Zoœcia*). "Every avicularium," he contends, "consists of a chamber, of variable size and shape, in which is lodged an apparatus of muscles, of a moveable horny appendage, which is worked backwards and forwards by the muscles, and of a fixed frame opposed to it, surrounding an aperture, upon which it falls when closed. In many cases, if not all, the chamber also contains a cellular body, which is in all probability the homologue of a polypide." It is hardly necessary to follow him through the various modifications of these appendages, although casually there would appear to be a very great difference, almost devoid of any similarity of appearance, between the lowest and most imperfectly developed forms, and the genuine bird's-head processes.

"The function of the avicularia," he says, "it is difficult to determine, nor, indeed, can the same function be assigned to all of them. The primary forms are many of them quite unfit for prehensile work. The

---

[1] Johnston, "British Zoophytes," vol. i. p. 333.

lid-like mandible, with plain rounded margin, has no power of grasping, and could not detain for a second the active worms which are sometimes captured by the articulated kinds. Their service for the colony must lie in some other direction. Even the fixed transitorial forms, in which the beak and curved mandible are present, must be inefficient for this work, from their want of mobility, whilst in many of them the parts concerned in the act of prehension are but slightly developed. The articulated avicularia, however, are undoubtedly grasping organs; and the presence of the tactile tuft, between the jaws, must be taken to indicate that capture in some form or other is their function. They have been seen to arrest minute worms, and hold them for a considerable time with a tenacious grip, as if with some ulterior object; but what the object may be it is difficult to decide. On the whole, I am inclined to regard the avicularia as charged with a defensive rather than an alimentary function. They may either arrest, or scare away, unwelcome visitors. Their vigorous movements, and the snapping of their formidable jaws, may have a wholesome deterrent effect on loafing annelids, and other vagrants; whilst the occasional capture of one of them may help still further to protect the colony from dangerous intrusion." The vibraculum, or moveable bristle, is related to the avicularium, and a line of

SEA-MAT MAKERS.

transition-forms link the two together. The setæ are sometimes of enormous size, and of great strength, and in certain species assume a locomotive function acting probably as oars, and propelling the colony which in that particular group are free in the adult state. This justifies the remark that "in the history of these appendages we have a curious illustration of the variety of function that may connect itself with the same morphological element."

Reproduction is secured in these Polyzoa by two methods, by regular sexual elements, and by the asexual method of budding, or gemmation. The majority of species are monæcious, male and female organs being present in each cell, but some species are unisexual, either wholly male or wholly female. A contractile cord, called the funiculus, is attached to the bottom of the stomach, and passes down to the bottom of the cell. It is considered an established fact that the male elements are derived universally from this funiculus, and in a great number of species the ova are also developed in the funiculus, sometimes apart from this organ, and sometimes attached to the cell-wall. It is still a question whether the ova are fertilised by male elements, developed in the same cell, or whether by those liberated from other cells. Their dispersion in immense numbers into the surrounding water seems to indicate the latter conclusion, and it is not improbable that in different species both methods may prevail. We cannot be expected to discuss at length, and in detail, a point of this character in a work like the present, but would refer those interested in the subject to the Introductory chapter of the volume already quoted.[1]

The ovary varies considerably in size; in some species it contains as many as thirty eggs, in others only one or two. At a certain stage of maturity they make their escape, through the ruptured wall of the ovary, into the space between the polype and the inner wall of the cell. After fertilisation the ova pass through the ordinary phenomena of segmentation, and are developed into free ciliated larvæ. In some species there is developed, during the breeding season, a small somewhat globose appendage to the mother-cell, called an ovicell, the interior of which is in direct communication with the internal cavity of the mother-cell. At first the ovicell is empty, afterwards it is occupied by the embryo. Huxley was the first to observe that impregnation takes place in the cavity of the cell, and then the egg passes from thence into the ovicell, where, as in a marsupial pouch, it undergoes its further development, and after undergoing yelk-division becomes a ciliated embryo.[2] Mr. Hincks

---

[1] Hincks, "British Marine Polyzoa." London, 1880.
[2] J. H. Huxley on "Reproductive Organs of the Cheilostome Polyzoa," in *Quart. Jour. Micr. Soc.*, vol. iv. (1856), p. 191.

adds, that "there can be no doubt that the ova generated in the mother-cell (*Zoœcium*) do pass into the ovicell, and there ripen into the perfect larva, escaping at last through the orifice. The ovicell is thus both a brood-chamber, and the passage for the embryo from the cell to the surrounding water. But whilst I have no doubt that the ovicell acts as a kind of marsupium, there seems to me to be grounds for believing that, in some cases and under conditions which I cannot explain, ova are also produced within it."

The larvæ, developed from the eggs, and which are the units of a new polyzoan colony, are very variable in form and complex in structure. They "are restless in their habits, and during their short term of free existence, are in almost constant movement—now whirling rapidly hither and thither, now tumbling over and over in the water, now creeping along, making use of their cilia as feet. Besides their ciliary appendages, they are often furnished with long setiform processes, which wave to and fro, and lash the water with much vehemence. After a while their energies fail, and they settle down and become attached; the cilia begin to flag in their movements, and soon disappear; and the volatile, and curiously organised being, resolves itself into a fixed and (apparently) homogeneous mass, in which the first polyp-cell (*Zoœcium*) and polypide originate."

Reproduction by gemmation, or budding, is constantly in progress, and by this means the plant-like forms of the erect branched colonies, the frond-like expansions of the sea-mats, and the flat patches of the encrusting colonies are formed. From the initial cell, which is the commencement of the colony, other cells are developed by budding, after the manner of the particular species, either longitudinally or at the margins, and, gradually, the colony enlarges and increases the outer cells, being the youngest, in a manner analogous to that in which the coral colony is increased. As we look upon a frond of the common sea-mat (*Flustra foliacea*) with its myriads of little cells (fig. 61), we may recall the fact that, once on a time, this frond was represented by a single mother-cell, and that, by continuous and consecutive budding, all those innumerable cells have been added, until the single cell has multiplied into a large colony, the polypes at the base having long since died, leaving

FIG. 61.—SEA-MAT (*Flustra foliacea*).

their empty cells as mementoes, whilst the marginal cells are not only living, but at the extreme margin still budding extremely juvenile cells, and thus the process continues to go on, so long as the colony, in its entirety, is in a living and thriving condition. In the encrusting species, many of them go on depositing a thick layer of calcareous matter, until the outer wall acquires much thickness and strength.

The latest enumeration gives no less than two hundred and thirty-five species as inhabitants of the British coasts, and some of these are extremely common. The dead skeletons, or tenantless homes, are cast upon the beach by every storm, and these objects must be familiar to every lounger on the shore. One of the largest, most plentiful (figs. 61, 62), and most popular is the common sea-mat (*Flustra foliacea*), which is sure to be picked up, with the residue of that odd miscellany

FIG. 62.—SEA-MAT (*Flustra foliacea*).

of mementoes, which every one brings home from the sea-side. Mingled with shells, bits of sea-weed, starfish, zoophytes, &c., is always the inevitable sea-mat, and yet how few trouble themselves about the animals that constructed those tiny dwellings, and inhabited them, increasing and multiplying till the little one became a thousand, and the single cell a great colony. It may be somewhat conducive to its popularity that it has a pleasant odour when fresh out of the water, which some compare to that of violets, others to bergamot, or the mixed odour of roses and geraniums. Sir John Dalzell states that he has known no less than ten thousand young embryos to have been liberated from a single specimen of this sea-mat, during the course of three hours. Growing upon, and attached to, this species, a much smaller one should be carefully sought for (*Bugula avicularia*), because it is often found in that position, and because it will exhibit those curious bird's-head processes, to which allusion has been made. They will be dead, as a matter of course, and all the snapping over, but the birds' heads will still remain, and imagination will readily picture them, in all the snapping energy of their original vitality.

Another characteristic Polyzoon will be found encrusting tufts of red sea-weeds, sometimes completely coating them, and bristling with long setæ (*Membranipora pilosa*) (fig. 63). " Its little cells have a metallic lustre, and almost look as if they were wrought in silver and richly chased." It is one of the most abundant of British Polyzoa.

Other encrusting species, belonging to different genera, will be found forming greyish, or whitish, patches on stones, rocks, shells, &c., but the description or enumeration of these belongs not to our province. We have indicated a wide field for investigation, and given an introduction to two hundred and thirty-five minute British animals that construct their homes in the sea, and all differing in some respect from each other in the form or decoration of the houses that they inhabit.

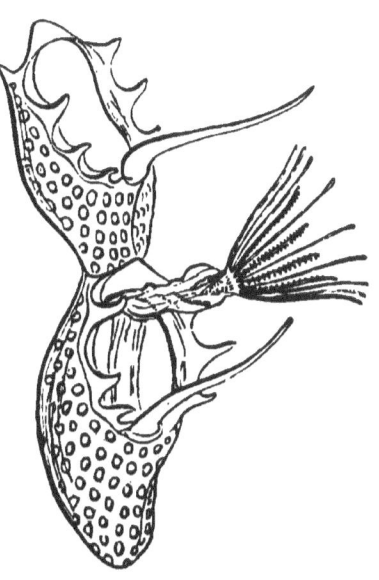

FIG. 63.—MEMBRANIPORA.

CHAPTER X.

TUBE-MASONS.

THOSE interesting insects, which are commonly known as Caddis-flies, make for themselves, when in the larval stage, minute tubes or cases, of many kinds of material, agglutinated together into the form of a tube ; sometimes it is chiefly of sand, or small fragments of shells, and sometimes of fragments of leaves, bark, or twigs. They are found in ponds, and slow-flowing canals, and the architects are larval insects of the order Trichoptera.[1] Analogous to these fresh-water cases we have also builders of marine cases or tubes, *not* the construction of insects, but of annelids or worms. In the caddis-case a temporary home is constructed, for a temporary purpose, and then abandoned, but the marine masons built their habitations for life, and do not usually quit them till they die. Nevertheless, with all their differences, there is an evident analogy between the

---

[1] See " Caddis-worms and their Cases," by R. McLachlan, F.R.S., in *Science-Gossip*, July, 1868.

case-building caddis larvæ of fresh waters, and the tube-masons, or tubicolous annelids of the sea. There are some, perhaps, who would call it mimicry on the part of one or the other, although unjustifiably; whilst some would blame us, on the other hand, for recognising any analogy between them. This is a point we are not anxious to discuss, although a comparison of the two forms of cases would be strongly suggestive, even if nothing more. The marine tubes, or cases, are constructed by annelids as a protection, and for the most part by a particular group of Tubicolous annelids, although even to this rule there are exceptions, for a few of the Rapacious annelids .(such as *Northia tubicola* and *Nereis diversicolor*) construct and live within a tube. Beyond these we are unacquainted with other tube-marine masons which have any claim to recognition in this chapter.

Sir Wyville Thomson mentions a curious little shell-fish (*Dacrydium vitreum*) dredged from a depth of 2,435 fathoms, "which makes and inhabits a delicate flask-shaped tube of foraminifera, sponge spicules, coccoliths, and other foreign bodies, cemented together by organic matter, and lined by a delicate membrane."[1] In colour it was a fine reddish-brown dashed with green.

---

[1] "Depths of the Sea," p. 465.

Certain tube-forming annelids (or worms) attach themselves to the corallum of living coral, and in some instances, so close is their relationship, that they have been called "messmates." For instance, in one species (*Mycedium fragile*) the smooth underside of the coral, which is destitute of coral polyps, is usually sprinkled with the tubes of a worm, in company with encrusting Polyzoa, and mollusks. The tubes of these worms are formed of carbonate of lime, like the coral on which it establishes itself.

"The calcareous tube, in which the worm lives, is firmly united through most of its length to the under side of the coral. Its terminal opening usually lies near the rim of the saucer-like disk of the young coral. In the growth of the rim this extremity of the worm-tube is imprisoned by the increasing edge of the disk, in such a way that it is wholly surrounded by the live coral in the progress of its growth. The growth of the worm-tube keeps pace with the advance of the coral about it, and, as it rises above the upper surface of the coral, the opening into the tube is kept uncovered by this growth, so that the head of the worm, with its crown of branchiæ, can always have free communication with the outside water, notwithstanding the tube itself is often wholly enclosed in the calcareous secretions of the polyps. When the growing coral covers the terminal opening of the worm-tube the worm dies, but this rarely occurs, as the closing of the opening is generally prevented by a corresponding growth of the tube. In several specimens examined the extremity of the tube had grown in a direction at right angles to the upper surface of the coral, and projects a full half inch above that plane. The coral, however, has formed along the sides of this projection, and has covered it to the very apex. The result is the formation of variations in the regular growth of the coral, which may be likened to coral gall, such as is formed by certain crustaceans. The greater part of the worm-tube lies in sight, on the under side of the coral, while its terminal opening is almost wholly concealed, being surrounded by the live coral community, through which there is, however, a small opening into the tube cavity, inhabited by the worm."[1] The modifications of the growing coral structure always commence in the vicinity of the opening of the worm-tube. As the growth of the coral proceeds, that portion which lies on the edge of the saucer-like disk surrounds the obstruction, and, when this takes place, the end of the worm-tube grows upward, and seems to rise out of the midst of the growing coral.

Other kinds of coral, belonging to other genera,

---

[1] "Annelid Messmates with a Coral," by I. W. Fewkes, in *American Naturalist*, vol. xvii. 1883, p. 595.

have also their companion tube-worms. Many specimens of these (*Porites*) have the interior of the coral mass riddled with these worm-cases, whose openings cover the surface of the coral head. " Such a combination of growing coral and worms, when both are alive, presents one of the most beautiful sights upon a live coral bank."

Mr. G. H. Lewes relates some curious experiences of one of the tube-making worms. " No one," he says, " I believe, has yet recorded the fact of the Terebella multiplying itself by the process of gemmation, which is known to occur in the case of some other worms. When the animal reproduces by this budding process, it begins to form a second head near the extremity of its body. After this head, other segments are in turn developed, the tail, or final segment, being the identical tail of the mother, but pushed forward by the young segments, and now belonging to the child, and only vicariously to the mother. In this state we have two worms and one tail. But in some worms the process does not stop here. What the mother did, the child does, and you may see, at last, six worms forming one continuous line, with only one tail for the six. The tail indeed is the family inheritance ; but reversing the laws of primogeniture, it always ·descends to the youngest, like that elaborate display of baby-linen, which was worked with such fondness for the first-born, and has become in turns the costume of successive pledges, as they appeared on this scene of life, with a constant diminuendo of interest, in all but parental eyes. Such, in a few words, is the budding of Annelids. The separation finally takes place, and then we perceive the children, and grandchildren, are not quite the same as their ancestor. The fact has not been observed at all hitherto in the Tubicolous worms, yet two of my Terebellæ gave me a sight of it. The first died before the separation took place. The second, after a day or two's captivity, separated itself from its appendix of a baby, and seemed all the livelier for the loss of a juvenile, which had been literally in the condition of "hanging to its mother's tail."[1]

We have already quoted from Mr. G. H. Lewes, and must do so again on the subject of the blood of worms. "That some animals have red blood, and others blood not red, you know perfectly well; but that the worms have blood of various colours is probably news to you. Thus the Sea Mouse has colourless blood; the *Polynoï* pale yellow; the *Sabella* olive green; and one species of *Sabella* dark red. But this difference

---

[1] Lewes, "Seaside Studies," p. 65.

of colour is trifling compared with the absence of corpuscles from the blood of all Annelids. The corpuscles, as you know, are the floating solids of the blood, and on them devolve the most important physiological functions; but the blood of all Annelids is entirely destitute of them.

"If we grant that the fluid hitherto universally regarded as blood is truly blood, we shall have to acknowledge that these Annelids have two different kinds of blood; for over and above the fluid, which we see circulating in the vessels, there is a fluid circulating, or, more correctly speaking, *oscillating*, in the general cavity of the body, and *this* fluid carries with it what are called the blood corpuscles. It consists of albumen and sea-water, and is named 'Chylaqueous fluid,' the simplest form in which blood makes its appearance, distinguished from the 'blood proper' in not being a fluid circulating in a system of closed vessels, but a fluid which carries the chyle directly to the tissues. An image may render the mechanism intelligible. Suppose a worm suspended in a phial of water. Let the worm represent the intestinal canal, and the glass phial represent the external integument, the water will then represent the chylaqueous fluid, which moves with every motion of the intestine, and fills up every cavity made by its motions." The albuminous and corpuscular nature of the chylaqueous fluid prove it to be subservient to nutrition.

"The two bloods have two methods of aëration. The 'chylaqueous fluid' rushes into the lovely tentacles, which in many species wave above the head, and there is aërated, aided by the action of the cilia, which line the inner surface of the tentacles. The 'blood' is carried to those arborescent tufts without cilia, which branch from each side of the head beneath the tentacles. But, although the respiratory process does undoubtedly take place in these organs, yet in animals so simply constructed, each organ performs more than one function."[1]

"The tentacles consist of hollow flattened tubular filaments, furnished with strong muscular walls. Each of these hollow band-like tentacles may be rolled longitudinally, into a cylindrical form, so as to enclose a semicircular space, if they only imperfectly meet. This inimitable mechanism enables each filament to take up and firmly grasp at any point of its length, a molecule of sand, or if placed in a linear series, a row of molecules. But so perfect is the disposition, of the muscular fibres, at

---

[1] See also Dr. T. Williams "On the Mechanism of Aquatic Respiration," in "Annals of Nat. Hist.," ser. ii. vol. xii. (1853), p. 395.

the extreme end of each filament, that it is gifted with the twofold power of acting on the sucking and on the muscular principle. In addition to the two important uses already assigned to these tentacles, they constitute also the real agents of locomotion. They are first outstretched by the forcible ejection into them of the peritoneal fluid, they are then fixed like so many slender cables to a distant surface, and then, shortening in their lengths, they haul forward the carcase of the worm."[1]

"The tubicolous Annelids, those modest recluses, who as soon as they emerge from the egg begin to construct for themselves a habitation, from which they never again depart. This habitation which is lengthened and widened according to the increasing bulk of the proprietor, is a tube, either calcareous or composed of a substance somewhat similar to leather, or wetted parchment. It completely envelopes the worm, which ascends and descends in the interior, without the necessity of rolling back its body (fig. 64), for its feet are constructed in such a manner that they can move backwards or forwards with equal ease and facility. These animals therefore pass their lives in a position somewhat similar to that of a child in swaddling clothes. The tube, which is

---

[1] Lewes, "Seaside Studies," p. 72.

closed at its posterior extremity, exhibits a circular opening in front, which serves as a kind of window through which these hermits are enabled to take a view of the world around them, to seize upon any prey which may happen to pass in their way, and to expose their blood to

FIG. 64.—TUBE-WORM (*Terebella conchilega*).

the vivifying action of the water, which serves them in the place of the air which we breathe. Do not, therefore, accuse them of curiosity or coquetry, because you see them so constantly display their richly-ornamented heads. But rather take advantage of this habit engendered of necessity, and carefully examine these marvellous forms. Do but drop into a basin of sea-water this fragment of a rock, and this old shell, whose surface is covered with serpulas and other tube-worms. Observe the prudent caution with which that little round plate rises above each tube, which it is designed to close hermetically, so that your eyes cannot penetrate to the interior. This is the shutter of the house; see, it is moving, the animal will soon show himself. Look, and you will see, below that operculum, bud-like patches of dark violet or rich carmine in one part, and of a blue or orange tint in another, while still further on appear tufts of every hue. See them expand, little by little, until they have displayed the whole of their thousand coloured branches, similar in form to a plume of ostrich or marabout feathers. You are a witness of the evolution of veritable flowers, more beautiful by far than the blossoms of our gardens, for these are living flowers. On the least shock, on the slightest shaking of the fluid, these brilliant petals close, and, disappearing with the rapidity of lightning, they retire within their stony tubes, whence they may defy their enemies from beneath the shelter of their operculum."[1]

One of the most remarkable of the tubeworms observed by Dalyell was called by him "the Weaver"

---

[1] Quatrefages, "Rambles of a Naturalist," vol. i. p. 108.

(*Terebella textrix*), which constructs a semicylindrical sand tube, of insufficient dimensions to cover the body, or receive the head. "A peculiar feature in its history is its producing a real cobweb, as distinct as that of the spider, with which it covers itself, and which also frequently, if not always, serves to support its spawn. The texture is very thin, rather irregular, and composed of the finest threads—almost invisible from their slenderness, and extreme transparence. Neither the mode of formation, or extension, nor the expedients for securing their extremities, are obvious. Such a web, from a specimen nine lines long, covered an area fifteen lines square. This is plainly the work of some exertion, as the threads, sometimes amounting to fifty, are fixed to the side of the vessel as high above the bottom as equals the length of the weaver, or more; and they also extend below, there to be secured. Thus it is evidently an artificial work, and it receives successive accessions. The specimen observed continued its work about three weeks in May, but although surviving a month longer, it wove no more."

Another Tube-forming Annelid (*Pectinaria belgica*) is found on sandy shores, within the lowest watermark, which "constructs a very delicate tube, as thin as paper, exclusively of the grains of sand agglutinated together in an extraordinary manner. The thickness of the side does not exceed a single grain;

each lies in its proper place, and the whole is lined with the slightest silken coating. The sand, being collected at the orifice of the tube, its tenant, chiefly by means of the tentacular organs, selects those which are appropriate, and applies them to use. This is done only through the night, all the additions being made at the orifice, and as the animal grows, the shape and dimensions of the tube result from the successive growth of the body."[1] Pallas says that the tube stands immersed in the sand, in a perpendicular position ; and indeed the worm is very helpless when it is laid horizontally. When at ease, and covered with the water, it protrudes from the wide aperture of the tube, the head with its four cirrhi, the comb of bristles, and its many tentacles. The latter are in continual movement ; they are shortened, and lengthened, and twisted about at will, in search seemingly for fit grains of sand ; and as the grains adhere, by a gluten secreted from the surface, they are carried within reach of the other organs, by means of which the worm applies them to the rim of its tube, and thus carries the structure upwards. The tube is only increased by addition to this end ; the posterior is plugged with the abdominal appendage, and undergoes no material alteration.

---

[1] Sir J. G. Dalyell, " Powers of the Creator," &c., vol. ii. p. 178.

The size of the tube corresponds exactly to that of the worm, and the animal can withdraw within it for shelter. It can also turn itself in the tube, so as to alter the relation of the back and belly to the sides of the case. The length of the worm is about two and a half inches.

Sabella and Serpula are names which are at least known to lovers of marine life, although the animals themselves are to many unknown. Sir J. G. Dalyell has given an interesting account of one species, which he says "is a timid, lively, active creature, whose most prominent property is constructing itself an artificial dwelling of the grains of comminuted sand, intermingled with shelly fragments, or other indurated substances. But there seems a great difference in the solidity of the dwelling, according to the position of the tube, or perhaps, the variety of the architects, which has never been the subject of sufficient observation. Thus we find the fabric, when a cylindrical segment running over some flattened surface, firm, durable, and capable of great resistance. It is not easily crushed. On the other hand, when cylindrical or alveolar, it appears to be always more brittle. Most of the dwellings of the *Sabella* are lined with a fine silky substance, formed of an exudation escaping from the body, which, consisting of indurated glutinous matter, is very conspicuous on breaking up the alveolar mass of some old congeries.

The animals testify a decided preference on choosing the materials of their habitations. While always preferring sand, and comminuted shell, pounded glass is sparingly and reluctantly employed, and unless for a few fragments, it is soon entirely rejected. But there is a striking difference in the character of the tubes. One is short and confined, extending little beyond mere accommodation for the body ; another is considerably prolonged, so as to afford a safe retreat in times of danger. The architect of a third seems to persist in advancing the fabric, as long as it can procure materials. It never wearies of working. Night is the chief season of architectural labour, though perfect idleness never leaves the day unoccupied. By means of the tentacular organs, and the cleft in the anterior part, grains of sand are selected and adapted to the precise spot, where glutinous matter secures them to the tube for sheltering its otherwise defenseless tenant."

The tubes of the scarcely common English species (*Sabellaria anglica*) form irregular masses, generally fixed amongst the branching roots of the large seaweeds (*Laminaria*) or heaped on old shells and stones. The size has no definite limits, and sometimes the tubes are solitary. In most cases the tubes are irregularly mixed, flexuous, an inch to an inch and a half in length, the spaces between them filled with sand, of the same kind as that of which the walls of the tubes are composed, the aperture, usually circular, a little expanded, and often tinged with purple.

The animal is thus described :—The body is less than an inch in length, of a somewhat quadrangular form, slightly tapered from the front to the tail, which is terminated with a narrow caudal appendage, usually curved, or bent upon the back; the foremost portion is generally coloured with purple, the abdominal region is straw colour, becoming pink or fine red at the posterior extremity. The head has the form of a circular disk, divided by a slit into two equal halves, and consists of three concentric rows of bristles, which have a pearly appearance. In the centre of the disk is the mouth, in the form of an elliptical fissure, surrounded with the inner row of bristles bent towards the orifice. The bristles are triangular, tapering upwards to a point, and striated crosswise. The bristles of the middle circle have a bulged, somewhat triangular head, supported on a narrow stalk. The marginal bristles point outwards each resembling a fork, with from five to seven unequal prongs, with the shaft flattened in the upper portion.

One of the commonest species of British Sabella (*Sabella penicillus*) has been well described by different authors. " The life of this species is an interesting history. Analogy leads us to believe that the eggs, involved in a mass of jelly, are extruded, in their season, from the upper aperture of the tube. Nourished in this jelly, they rapidly pass through the fœtal metamorphoses, and attain the annelidan form so soon as they are free, or, at least, before they are more than two lines in length. The first instinct constrains and enables the infant worm to build, out of the mud around it, a tiny case to shield the body; and this is in future always carried onwards as an advanced work, so as rather to receive the body, as it grows, than to wait upon that growth. The growth is rapid; and the external organs, as well as the rings of the serpentine abdomen, are involved in succession. Thus the worm has at first few segments, and the ornaments of the head may have no more than six filaments. These two are, on their first appearance, simple; and it is a subsequent development that fringes them with cilia, and makes the organ pectinated. When mature a well-fed specimen may have ninety-two filaments in each of the fans of the branchial tuft that adorns the head; and the body may consist of not fewer than three hundred and fifty rings, each with their pair of pencilled feet, and ringlets of many hooklets. Such a specimen will be fifteen inches in length, and the tube will be two feet."

" The tube is essential to the existence of its tenant. The part first formed, to fit it for its purpose, is necessarily of small calibre, and is laid horizontally. The inmate protrudes its posterior extremity from beyond the lower end of the tube, and fixes it to any adjacent object, or support, by a glutinous secretion, furnished apparently by the anal segment. When the tube has thus been anchored, the entire attention of the worm is directed to its anterior end, which is raised up, in a more or less erect posture, by continual and incessant additions to the rim of the aperture. To catch and collect the muddy material necessary for the work, the branchial fans are spread out into a semicircle, so that, when the two are brought into contact, a wide funnel is formed. Once in the funnel the muddy water is forced down the rachis of the filaments, by the play of the ciliary fringes, and brought within reach of the singular organ at the base of the funnel, by which the mud is selected and applied, just as a mason would lay lime on with his scoop, and then mould and smoothen it with his trowel. At the base of the funnel, and towards one side, are two external fleshy lobes, or trowels, with an organ like a tongue or scoop between them. Receiving the pellets of mud, the creature mixes it up with an adhesive secretion, furnished probably by the collar of the cephalic segment, and by the organs just mentioned. It is thus rendered consistent and tenacious, and fit to be employed in raising the edge of the tube. To that position the material is raised by the tongue and trowels, aided by a general elevation of the head; and it is fashioned into shape by the same tongue and trowels, curved over the exterior circumference, as far as they can be stretched, and smoothed and polished by their motions, while clasping it with their pressure."

"And thus the tube is built up. The lower portion has been left unoccupied, for it has become too straight for the tail, which has grown with the worm's growth; and the upper portion extends far beyond what may at first seem necessary, but its Creator foresaw that it was needful this lower work of His should be able at pleasure to hide the glories with which He has adorned it,—otherwise too seductive to the enemies that were enticed by the richness of the display. For its tube the worm seldom uses other material than soft mud, but in urgent need fine sand may be partially resorted to. The gummy fluid, with which it is cemented, is, in the first instance, undoubtedly supplied by organs connected with the head; but much is afterwards furnished by the skin of the body, to make the interior more consistent and lubricous. Indeed, that the tube may be kept circular throughout, the worm is, while working, in a state of continual rotation."[1]

---

[1] "Johnston's "Catalogue of British Worms," p. 257.

If a glass jar containing this Amphitrite (as he calls it) be emptied and replenished with sea-water, Sir J. G. Dalyell says that the animal will retreat, the orifice of the tube be closed, and all at rest. "But soon after replenishment it rises to display its branchial plume still more vigorously than before, and remains stationary, as if enjoying the freshness of the renovated element, always so grateful. The passing spectator would conclude that he now beholds only a beautiful flower, completely expanded, inclining towards the light, like some of those ornaments of nature, decorating our gardens. He pauses in admiration. But if a drop of liquid mud falls amidst the element, from above, disturbing its purity, then, while the plume unfolds to its utmost capacity, does the animal commence a slow revolution, the body also passing round within the tube. Now are the thousands of cilia fringing the ribs of the branchiæ discovered to be in vigorous activity, and their office to be wondrous. A loose muddy mass is soon afterwards visibly accumulating in the bottom of the funnel; meantime the neck, or first segment of the body, rising unusually high above the orifice of the tube, exhibits two trowels beating down the thin edge, as they fold and clasp over the margin, like our fingers pressing a flattened cake against the palm of the hand. During these operations, muddy collections are seen descending between the roots of the fans towards the trowels, while another organ, perhaps the mouth, is also occupied, it may be in compounding the preparation with adhesive matter. Still does the partial or complete revolution of the plume above, and of the body within the tube continue; the bulk of the muddy mass diminishes, activity abates; it is succeeded by repose, when the tube is found to have received evident prolongation."

There is no permanent or organic connexion between the worm and its tube, but the worm never leaves it, except under circumstances which threaten a slow death. Removed from the tube by accident, the tenant cannot reoccupy it, nor reproduce another to cover its nakedness. But it lies exposed, ill at ease, and incapable of any motion beyond a painful writhing; casts off its ornaments; becomes weaker, and dies gradually downwards,—that is, the lower portion of the body retains life for some time after the upper has been dead, and begun to decompose. But so long as it remains within its tube there is no worm more tenacious of life, nor more richly endowed with the power of repairing wounds and losses. The beautiful tufts on the head are occasionally cast away, as if they were merely parts of its holiday attire; and the worm replaces them, with marvellous promptitude, with a pair not inferior to the first in beauty. A large proportion of the body may be amputated, and it will be restored ultimately to entirety ;—and, what is truly wonderful of a creature so complicitly and curiously made, a small portion of the abdominal part has grown to be a perfect individual, having reproduced segments of its own kind, thoracic segments of a different character, and the head and all its garniture and bravery.

Another tube-worm (*Protula protensa*) constructs a cylindrical tube from four to five inches long, and as thick as a goosequill, of an opaque chalky whiteness. The head of the animal is furnished with a pair of fan-shaped branchiæ, with its yellow filaments spotted with scarlet. Dr. Johnston kept one of these worms in confinement for several days to watch its movements. "The worm," he says, "would sometimes remain for hours concealed in its shell, and, when it ventured to peep out, the branchial tufts were sometimes slowly and cautiously protruded, and sometimes forced out at once, to their full extent. After their extrusion, they were separated and expanded, and lay at perfect rest on the bottom of the plate, in unrivalled beauty, and an object of never-failing admiration. The worm, however, seemed never either to slumber or sleep ; for, on any slight agitation of the water,—occasioned, for example, by walking across the room, or leaning on the table,—it would at once take alarm, and hurriedly retreat within the shelter of its tube. It was never off its guard, and would often, when lying apparently in calm indulgence, suddenly withdraw, in evident alarm, without a cause, but what was gendered by its own natural timidity; for the phantoms of dreams are not, it may be, the visitants only of higher intelligences, but come as they like, in a fearful or cheerful mood, even to these lower things. It never protruded itself far from its tube, and after becoming weak and sickly, it first threw off one half of its pride, a branchial tuft; and after several hours the other was likewise cast away, when the poor mutilated creature buried itself, still living, and to live for a day or so longer, in its own house and cemetery."[1]

Examples might be multiplied of these tube-forming annelids, but, with the exception of two well-known species, enough will have been written to demonstrate that they have a considerable interest of their own, apart from their constructive capacity, although they do not appear to have been made the subject of investigation by many of our own naturalists, which doubtless they would amply repay.

The Serpula, which is best known on our coasts (*Serpula vermicularis*), constructs a pinkish tube of

---

[1] Johnston's "Catalogue of Worms," p. 266.

about three inches in length, often attached to old shells, and differs from the other tubicolous annelids enumerated, by the possession of an operculum, with which to close at pleasure the orifice of its tube. This operculum is recognised as a modification of one of the tentacles, the other remaining in its normal condition. The tubes are homogeneous, and of a shell-like substance, sometimes solitary, but often clustered, and intertwined in a contorted manner (hence it has been called *Serpula contortuplicata*), wrinkled transversely, rounded, with a sharp dorsal keel, which is sometimes almost or entirely obliterated. The branchial tufts, and also the operculum, are blotched with a beautiful scarlet colour. In the various species there will be found to be considerable variation in the form of the "stopper," or operculum (fig. 65); in some species it is discoid, whilst in others it is funnel-shaped, almost like a miniature wine glass, and with the margin either even, or crenate, or lobed, or elegantly toothed, whilst in others it is complex, with two or three projecting ridges or galleries superimposed. The form and character of the operculum has been regarded by some as of primary importance in the classification of genera and species.

Fig. 65.

Operculum of Serpula intricata after Johnston.

z

Most common of all, is the Spirorbis (belonging to several species, but usually *S. nautiloides*), the little tubes of which, without their tenants, may be seen dotted over nearly every stranded frond or tuft of sea-weed, or dead shell. The tube is chalky white, and coiled round in a spiral, so as to resemble the shell of a tiny mollusk, and its whole diameter scarcely exceeds that of the head of a good-sized pin. If found, whilst still living, the small tuft of branchiæ, and tiny operculum, may be distinguished by means of a pocket lens.[1]

Many persons have a feeling of repugnance towards worms, and by them it is possible that the perusal of the present chapter may be omitted. And yet there are more marvels, and mysteries, associated with these despised creatures than we have hinted at. The microscopist, pure and simple, with no special proclivities, will find the study of the bristles an occupation of considerable time and interest. Johnston has devoted several pages to them in his volume, and these merely suggestive, and not exhaustive, which he concludes thus :—" Let me ask the naturalist to count the number (of bristles) which may be required to furnish the garniture of a single individual. There are annelids which have five hundred feet on

---

[1] See Dr. Philippi, " Observations on the Genus Serpula," in "Annals Nat. Hist.," 1st series, vol xiv. (1844), p. 153.

each side,—each foot has two branches,—and each branch has at least one spine, and one brush of bristles, some of them simple, some of them compound. This individual then has 2,000 spines at least; and if we reckon ten bristles to each brush, it has also 20,000 of them! This, as Sir Thomas Brown would say, is one of the 'magnalities' of nature; yet, let us look a little further, not merely to the exquisite finish of each bristle, but to the means by which the host is put in motion. There is a set of muscles to push them forth from their port-holes; there is another to replace each and all of them within their proper cases; and the uncounted crowd of these muscles neither twist nor knot together, but play in their courses, regulated by a will that controls them more effectually than any brace,—that now spurs them to convulsive energy, now stills them to rest, and anon puts them into action, where the ease and grace charm us to admiration, and fix the belief that even these creeping things participate largely in the happiness diffused throughout creation!" Hush! reader, it is only a worm!

CHAPTER XI.

EXCAVATORS.

EXCAVATORS, or boring animals, are some of them so destructive, in the pursuit of their legitimate occupation, that they have obtained an unenviable notoriety amongst men, who are often scheming to circumvent them in their nefarious designs. Although somewhat beyond the scope of our original design to include the marine mollusks within our category, yet, as a preliminary to the more minute and insignificant excavators, it will be necessary to devote a few lines to the boring mollusks. Foremost, perhaps, in notoriety will be the Teredo, that terror of shipowners, and those interested in wooden structures exposed to the sea. These creatures have the audacity to attack every piece of wood which comes within their reach, perforating it in all directions, until at last it crumbles at a touch. Many a ship has split in the open sea, through the planks having been drilled by this insidious invader. The hardest oak or teak wood is no obstacle to its depredations. "The animal always tunnels in the direction of the grain, and if in its course it meets with another, engaged in the same process, it alters the direction of its course, so that a piece of wood attacked by many Teredos becomes completely honey-combed." In the beginning of the century, half the coast of Holland was threatened with the invasion of the sea, because the piles, which upheld the dykes, were attacked by the Teredo ; and it required an outlay of a large sum of money to secure the country from the disaster of an inundation, caused by a contemptible mollusk.

"The body of the Teredo is long and worm-like ; its colour greyish-white (fig. 66). At one end is a knot, improperly called its head, and the other extremity bifurcates into what may be called two tails. It often attains the length of eight inches. It buries itself in a case, which it eats out of the wood ; the walls of the case are covered with a coating of a calcareous matter mixed with mucus ; this renders them firm and solid. The round part of the mollusk carries two small valves, very thin and fragile ; they are not unlike in shape to two halves of a

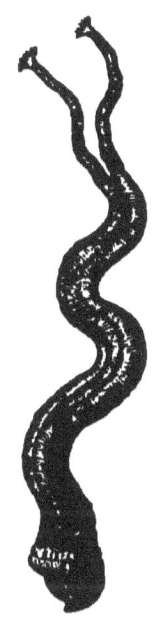

FIG. 66
(*Teredo navalis*)

nut-shell. These valves are immoveable, and protect the weak part of the animal. All the internal organs of the Teredo are placed one immediately behind the other, a position which is necessitated by the narrowness of the body, and the repeated elongations which it is required to undergo."

"The Teredo deposits greenish-yellow eggs, which are spherical. As soon as the larva is born, it becomes covered with vibratory cilia. It swims with facility, rises and falls in the water, seeking for wood into which it may penetrate. When it has found a piece of wood to suit its convenience, it walks up and down on its surface, like a caterpillar. From time to time it opens its little valves, as if practising for the mining operations its contemplates. At last it fixes upon a place, and makes an incision. In due time it has excavated a hole, capable of containing half its body. The young Teredo now covers itself with a coating of mucous matter, which, condensing by degrees, at last forms a resisting shell. In this covering two or three holes are left, for the projection of the syphons. About the third day the tube has become solid, and it is now the origin of the tube of the animal. We cannot see what is taking place inside, for this coat is opaque; but if we detach some of the young Teredos from the wood at this stage of their development, we shall find that there has been secreted a new shell, similar in all respects to that of the adult animal. It cannot be that this shell is the instrument wherewith the perforation is executed, for it is very fragile; and it would surely show some signs of wear and tear, which is never the case."[1]

It has been the source of much speculation, and inquiry, how these animals, without powerful boring organs, can make their tunnels in the hard wood. Some have imagined that the animal secretes some fluid which corrodes the wood, so that it may easily be removed. Yet no one has been able to prove the existence of such a corrosive fluid in connexion with the animal. Professor Owen has explained that it needs no corrosive fluid, but that the muscular action of the mollusk is in itself sufficient.

By whatever method the boring may be accomplished, it is of more importance to discover how some check to their ravages may be devised. M. de Quatrefages has shown, by experiment, that this feat may be accomplished. After they have attained their adult state the Teredos live in secluded galleries, hence some have imagined that each individual is at the same time male and female, whereas this is not the case, for they are of

---

[1] Moquin-Tandon's "World of the Sea," p. 193.

different sexes. At a certain time the eggs are emitted by the female, and rest within the folds of the respiratory organs, awaiting fertilisation by the male element. This latter is diffused promiscuously in the water, " the organic particles, disseminated through the surrounding water, are carried away by the currents, and some of them thus make their way into the respiratory organs (branchiæ) of the females, where they meet with the eggs, and vivify them by contact. If we could only destroy the vitality of these organic particles, before they come in contact with the eggs, we could at once put a stop to the further development of these destructive mollusks. If a twenty-millionth part of a mercurial salt be thrown into the water, in which myriads of these organic corpuscles were moving, it would be sufficient to render all of them motionless, in the course of two hours. A two-millionth part of the same salt produces the same effect in forty minutes, and entirely deprives the water of the fertilising power, with which it had previously been so highly charged. To preserve the submerged wood of our marine docks, and wharfs, it is therefore only necessary, from time to time, to throw a few handfuls of these different substances into the surrounding water. Fecundation would thus be entirely arrested, and the eggs would perish before they were developed, and consequently the species would be completely exterminated, in our harbours and docks, in the course of two or three seasons."[1]

Another mollusk excavates not only wood but stone (fig. 67). It is the *Pholas*, of which there are several species. The pair of shells are very thin and delicate, gaping at both ends, and when applied to the animal, with the syphons extended, has an elongated conical form, which is doubtless very useful in its boring operations. Although living usually in stone or wood, they may be detached, and examined "all alive," with the pair of syphons extended at the narrow end. The water enters by the larger syphon, and is expelled again from the smaller. The lining membrane is coated with vibratile cilia, which, by their incessant action, keep up the current of the water. A great deal has been written on the subject of how its perforating operations are performed, which resolves

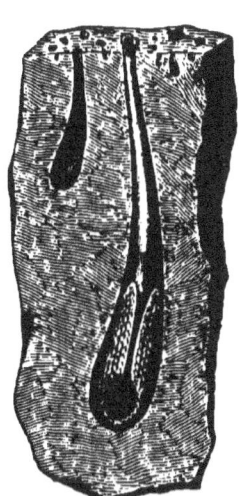

FIG. 67 (*Pholas dactylus*).

---

[1] Quatrefages, "Rambles of a Naturalist" (London, 1857), vol. ii. p. 231. The author states that one pound of corrosive sublimate or two pounds of "sugar of lead" would destroy all the organic corpuscles in more than twenty thousand cubic yards of water.

itself at last to a very simple process, of continuous motion in a given direction, without any assistance from a supposed acid secretion, which earlier authors considered necessary to dissolve the rock operated upon (fig. 68).

There is also another excavating mollusk (*Xylophaga dorsalis*) which causes considerable damage to timber work about docks, and to that of piers and jetties. It attacks timber of all kinds that are under water in the neighbourhood of quays. " Like the *Teredo*, it inhabits the interior of wood, which has been some time under salt water, penetrating to the depth of from half an inch to an inch, forming for itself an oval receptacle or cavity, and having a very small and single external orifice."[1] It differs

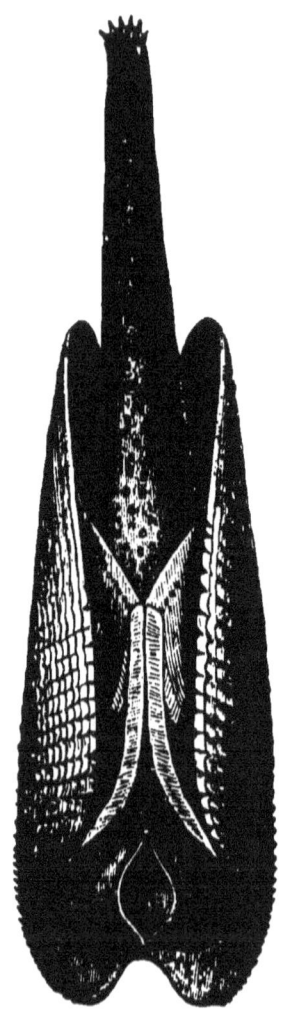

FIG. 68 (*Pholas dactylus*).

[1] Mr. W. Thompson " On *Xylophaga Dorsalis*," &c., in " Annals of Nat. Hist.," vol. xx. (1847), p. 157.

from the *Teredo* in that it bores against the grain of the wood in a diagonal manner. The perforations of the two species may be observed in the same piece of timber, for they seem to labour harmoniously together in the work of destruction. Many of the chambers have been found an inch and a half in length, whilst the largest shells observed are five and a half lines in length.

Dr. G. C. Wallich has also borne testimony to the boring operations of minute Annelids at great depths in the sea, for he records that he "has met with several examples of Foraminiferous shells, brought up from the greatest depths, perforated, in all probability, by the minute boring Annelids that construct and inhabit the tubes of which he had made mention. The extreme delicacy of the inhabitants of these tubes has, as yet, completely baffled him in all his attempts to extract them, and determine their character. In addition, however, to the tubes, formed in so singular a manner, of innumerable Globigerina shells cemented together, there also occurred other tubes, in which the internal layer was a cylinder of tough membraneous material, with a rich sienna tint, whilst its outer surface was strengthened and protected, partly by numerous Globigerina shells, as in the previous case, and partly by a layer of siliceous spicules, probably derived from some minute sponge. The perforations in the shells were invariably of one character, and consisted of an aperture bored through and through, but having the entire thickness of the shell wall, from the inner surface to the outer one, as it were countersunk. Accordingly, in section, such a perforation would present a truncated cone, the apex of which is directed inwards."[1] These observations seem to indicate that double-capacity, exhibited by many annelids, of boring chambers for themselves in hard substances, and then lining them with a tube, or of constructing independent tubes, and thus being either excavators, or tube masons, according to circumstances.

There is a brown or dusky worm, about an inch in length (*Dodecaceria concharum*), which bores its chambers in one of the hardest of our marine shells. It has no eyes and no proper head, but there are a pair of long tentacles on the terminal joint, which occupies the place of a head, and beneath these three or four pairs of slender filaments. "When at rest under water the worm protrudes the tentacles and filaments from the circular opening of its burrow. The filaments are laid along the shell, and kept quiet, or moved about like independent worms." Johnston never saw them capture any prey. "The excrements are pushed out at the same aperture, and may be

---

[1] "On the Boring Powers of Minute Annelids," by G. C. Wallich, in "Annals Nat. Hist.," vol. ix. (1861), p. 57.

seen occasionally collected in small earthy pellets about the margin."[1]

Singularly enough, there are at least two kinds of excavating crustaceans (allied to the shrimps) which are found around our shores, excavating timber in the sea, destroying the woodwork of piers to a most serious extent. At first called the *Limnoria* and the *Chelura*, names since amended and altered, we shall continue to designate them as such. Both the crustaceans labour harmoniously together in the work of destruction, and are mingled in the wood as if they were all of one species. " They can be readily distinguished from each other, either when alive or dead, the *Chelura* being of a reddish, the *Limnoria* of a pale greyish-yellow hue, resembling that of light-coloured pine or fir. As they retain their colours after death, we may even years afterwards distinguish the two species in the excavations which they had formed, in timber subjected to their ravages. From this circumstance, added to that of their burrows being formed in the closest contiguity, and many of the creatures dying in them, after the timber has been removed from the sea, we may in our museums display whole catacombs of them, as closely packed as ever were mummies in the best-

---

[1] "Catalogue of British Non-Parasitical Worms," by George Johnston, M.D. (London, 1865), p. 214.

tenanted tombs of Egypt."[1] Professor Allman says that "it excavates the timber not merely for the purpose of concealment, but with the object of employing it as food, which is apparent from the fact that the alimentary canal may be found on dissection filled with minutely comminuted ligneous matter. Timber, which has been subject to its ravages, presents a somewhat different appearance from that which has been attacked by *Limnoria*. In the latter we find narrow cylindrical burrows running deep into the interior, while the excavations of *Chelura* are considerably larger, and more oblique in their direction, so that the surface is rapidly washed away by the action of the sea."

We have already quoted the authority of Professor Agassiz for the destructive action of excavators on the old masses of dead coral, and we will again refer to his remarks in this connexion, as they show the extensive character of these operations. "Innumerable animals," he says, "establish themselves in the lifeless stem, piercing holes in all directions into its interior, like so many augers, dissolving its solid connexion with the ground, and even penetrating far into the living portion of these compact communities. The number of these boring animals is

---

[1] W. Thomson "On *Chelura Terebrans*," &c., in "Annals Nat. Hist.," vol. xx. (1847), p. 161 ; and Prof. Allman in "Annals Nat. Hist.," vol. xix. (1847), p. 362, with figures.

quite incredible, and they belong to different families of the animal kingdom. Among the most active and powerful we would mention the date-fish, and many worms, of which the *Serpula* is the largest, and most destructive, inasmuch as it extends constantly through the living part of the coral stems. On the loose basis of a brain-coral, measuring less than two feet in diameter, we have counted not less than fifty holes of the date-fish, some large enough to admit a finger, besides hundreds of small ones made by worms. But however efficient these boring animals may be in preparing the coral stems for decay, there is yet another agent perhaps still more destructive, we allude to the minute boring-sponges, which penetrate them in all directions, until they appear at last completely rotten through." This last remark, as to the boring sponges, should be borne in mind, coming as it does from such an authority, when we advert to the diversity of opinion, still existing, as to the capacity of sponges as excavators.

The boring annelids, or worms, have never had their penetrating powers called in question, so that it may be conceded that they are ever industriously at work, in drilling their sinuous galleries, not only through the dead stems of coral, but also into the hard rock. They belong to numerous species, some of them classed with the tube-forming annelids, from their habit of lining their excavations, so as to construct buried tubes. In structure they correspond in all respects with their congeners of tube-making propensities, and hence will require no special description. Doubtless, they are not by any means guiltless of perforating old shells, and living ones too for that matter, although it is by no means proven that the chambered oyster-shells, lined with sponge, have been at first perforated, as some assert, by small annelids, and afterwards occupied by the sponge. It is difficult to conceive how all the dendritic cavities in shells were excavated by worms, whilst, it must be confessed, that certain forms of excavation bear strong evidence in themselves of their annelid excavators. Direct proofs of such operations are difficult to obtain, inasmuch as they are performed far beyond the reach of prying eyes, and hence much must depend on inference from analogy. Whether there is any real analogy between the operations of a boring annelid, or a boring sponge, and the larvæ of a wood-boring beetle (*Scolytus*), we will not pronounce, although at least one writer seems to infer that what is true of one must be true of the other. There is at least sufficient resemblance, in the results, to make the gradation easy from the worm to the sponge, and enable us to give here, in fuller detail, what has been affirmed of the latter, which, as Agassiz says, are " perhaps still more destructive " than the former.

The boring sponges (*Cliona*), are a group of exca-

EXCAVATORS.

vators, of very common occurrence, even in our own seas. Nearly every oyster-shell, which is cast upon

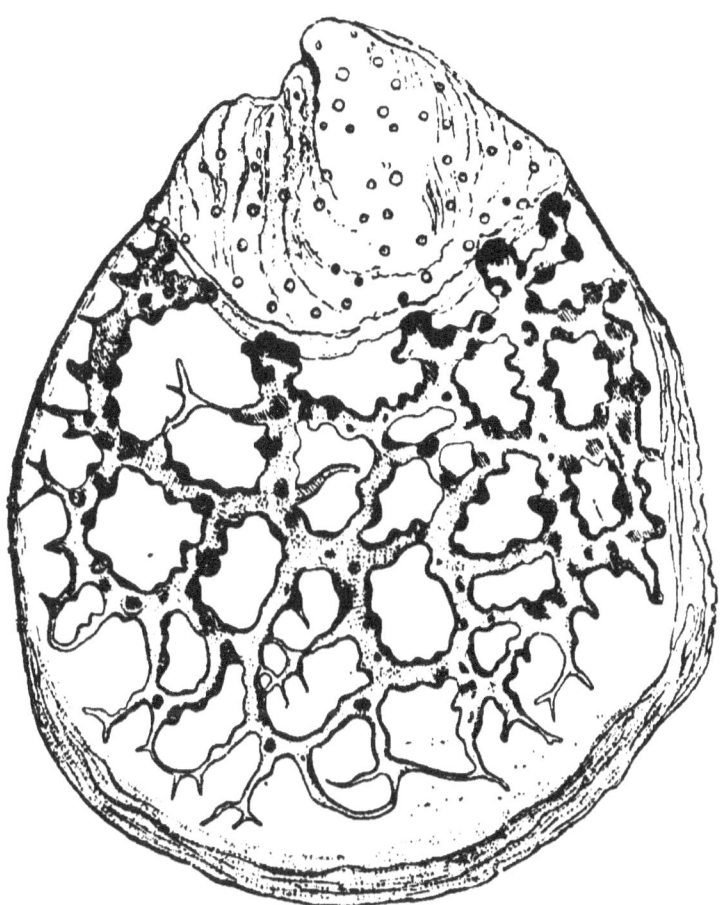

FIG. 69.—SHELL WITH BORING SPONGE.

the beach, is found to be perforated and chambered by some mysterious agency, and the cavities filled with a species of sponge, found only in such a position (fig. 69). Whether or not the sponge itself has bored the holes, and excavated the chambers, some have ventured to doubt, but that the sponge called the " Boring sponge " or " Burrowing sponge " is always found lining these cavities, is sufficient to justify us in treating it as an excavator. Dr. Leidy gives a full and particular account of the sponge in its living state. "This boring sponge," he says, "forms an extensive system of galleries between the outer and inner layers of the shells, protrudes through the perforations of the latter tubular processes, from one to two lines long, and one-half to three-fourths of a line wide. The tubes are of two kinds, the most numerous being cylindrical, and expanded at the orifice, in a corolla form, with their margin thin, translucent, entire, veined with more opaque lines, and with the throat bristling with siliceous spicules. The second kind of tubes are comparatively few, about as one is to thirty of the other, and are shorter, wider, not expanded at the orifice, and the throat unobstructed with spicules. Some of the second variety of tubes are constituted of a confluent pair, the throat of which bifurcates at bottom. Both kinds of the tubes are very slightly contractile, and, under irritation, may gradually assume the appearance of superficial, wart-like eminences, within the perforations of the shell occupied by the sponge. Water obtains access to the interior of the latter through the more numerous tubes, and is expelled in quite active currents from the wider tubes."[1] The general sponge structure has already been detailed in the chapter on sponges, and needs not repetition.

Notwithstanding the doubts, which some writers have expressed, of the possibility of sponges boring into hard structures like coral, or the shells of mollusks, we submit a few facts for consideration on the other side of the question. " In 1871, a vessel laden with marble, was sunk in Long Island Sound, and, according to Professor Verrill, the boring sponge had penetated the exposed parts of the blocks (in 1879), for a depth of two to three inches from the surface. The canals vary from one-fourth to a hundredth of an inch, and less, in diameter ; the canals are coated within with a thin film of dried sarcode of a brown colour, which was orange coloured in life. Though the sarcode is dried, the needle-shaped spicules are plainly visible, under a one-fifth inch lens, and display the form usually seen in the same species found on the coasts of Europe. The specimen, which I have seen, shows a series of large branching canals, which connect freely with each other, in the most irregular way imaginable ; moreover, the form of the canals, in transverse section, is exceedingly variable, being

---

[1] "American Magazine of Natural History," vol. i. (1878), p.54.

oval or irregular as often as it is circular. These last facts, together with that of the great variability in the calibre of the canals, leaves no doubt in my mind that it is the animal of the sponge which does the boring, and not marine worms, which have politely abandoned their burrows for the accommodation of this toiler in the sea."[1]

Dr. O. Schmidt, who may be supposed to have had a rather intimate knowledge of sponges and their habits, observes that "a large portion of the coasts of the Mediterranean and Adriatic Seas is composed of calcareous material, which, from its tendency to become eroded, has a broken, jagged aspect, giving it a peculiar and often attractive appearance. Of such broken coast, one can certainly measure off some thousands of miles of strand, and, where it does not descend too abruptly, large and small stones and fragments of rocks cover the ground. One can scarcely pick up one of these billions of stones, without finding it more or less perforated with holes, and eroded by *Cliona* (burrowing sponge), often to such a degree that the spongy remains of the apparently solid stone may be crushed by the hand."

Again, Dr. Leidy remarks that "an extensive bed of oysters, which had been planted at Great Egg

---

[1] "Destructive Nature of the Boring Sponge," by J. A. Ryder, in "American Naturalist," vol. xiii. (1879), p. 279.

Harbour, and which was in excellent condition three years previously, had been subsequently destroyed by an accumulation of mud. The shells of the dead oysters, which were of large size, and in great number, in the course of two years had been so completely riddled by the boring sponge (*Cliona*) that they might be crushed with the utmost ease, whereas without the agency of this sponge, the dead shells might have remained in their soft, muddy bed, devoid of sand and pebbles, undecomposed, perhaps, even for a century." [1]

In reference to the mode by which this burrowing is accomplished, Dr. O. Schmidt says :—" One would first think of the siliceous needles as the cause, but we soon see that we must abandon the notion that this is the boring apparatus, since it must be borne in mind that such apparatus must be operated. Even though the protoplasm executes delicate fluctuating movements, so that in *Cliona*, as in many other sponges, the needles are drawn into bundles, rows, or series in particular directions ; in any case, the force so exerted would not be sufficient to scrape or erode the lime rock with their points. This mode of distribution and extension of the sponge would rather indicate that a process of chemical solution

---

[1] Leidy, in " Proceedings of Academy of Natural Sciences of Philadelphia," vol. viii. p. 162.

was the real agent at work in erosion. Of the exact constitution of this corrosive fluid we know nothing. The importance of the boring sponge in helping to effect the redistribution of eternal matter, does not consist in comminuting the stone into particles, but in dissolving it, as sugar is dissolved in a glass of water, and mingled with the sea-water in this dissolved condition. Out of this solution the innumerable shell-fish take the mineral materials, which have been mingled with their blood, and from which it is deposited as new layers on the shell, which, when the animal dies, either is also finally re-dissolved by the sponge, or falls to the bottom of the sea, as a contribution to the earth's strata of future æons."

The most explicit, and detailed account, which we possess of these burrowing sponges is that of Mr. Hancock, in 1849, in which he strongly insisted on the excavations being performed by the sponges themselves. Although, at the time, his conclusions did not meet with universal assent, his case was evidently a strong one in favour of his hypothesis. "On the coast of Northumberland," he says, " the surface of almost every piece of limestone, near low-water mark, is riddled by *Cliona* (boring sponge) ; old shells, whether univalves or bivalves, are filled with it ; it inhabits millepores ; and in southern latitudes it buries itself in corals. Its ravages are very extensive, and appear to be rapidly effected. I have seen half-grown living oysters, with *Cliona* extending from the umbones, almost to the ventral margin, and in one or two instances it even reaches the margin. In these cases it is evident that the growth of the sponge must have been more rapid than that of the shell; for the work of destruction could not commence until the oyster had attained to some size; and had its growth been even equal to that of the sponge, the shell ought to have reached its full development, before the sponge had gained the lower margin. When a shell is once attacked, the operations of these creatures never cease, until they have extended throughout its entire substance. The middle portion soon becomes almost completely excavated, small pieces only remaining to divide the chambers or branches. A thin plate is left on the outer and inner surfaces to protect the parasite; and even these plates are ultimately riddled with numerous circular holes, which are the only indication of the work of destruction beneath, until some slight external influence ruptures the protecting walls, or the increasing growth of the tenant bursts them asunder; when the whole system of elaborately-wrought chambers becoming exposed soon give way, and *Cliona*, Sampson-like, perishes amidst the ruin produced by its own energy."

The description, which he afterwards proceeds to give, of the channels made in the shells indicates one of the grounds upon which he attributed their formation to the sponges themselves. "The boring sponges, as far as I have examined them, are branched, or are composed of lobes united by delicate stems, and all more or less anastomose, according to the species ; many of them are beautifully arborescent, and of great delicacy. They all bury themselves in shells, or other calcareous bodies, and communicate with the water by papillæ, or oscula, protruding through circular holes, in the surface of the containing substance or matrix. In dead shells the papillæ pass through both surfaces, but in living ones rarely penetrate the innermost layer, though occasionally they do so. When a mollusk is thus wounded, it deposits calcareous matter over the orifice, and generally succeeds in excluding the intruder. The species vary considerably in form, and might be divided into two or three distinct groups. In some, the branches are almost linear, and anastomose only to a very slight degree, others form a complete network, with the meshes so small that very little of the matrix remains between the branches ; some have the branches moderately lobed, and others, again, have the lobes large, and crowded upon each other in all directions, and united by fine, very short stems. In most the terminal twigs are very minute, and exhibit in a decided manner the mode of growth (fig. 70)." He then details the peculiarities in different species, and adds:—"Thus is the form of the excavating sponges varied, and the chambers they inhabit modi-

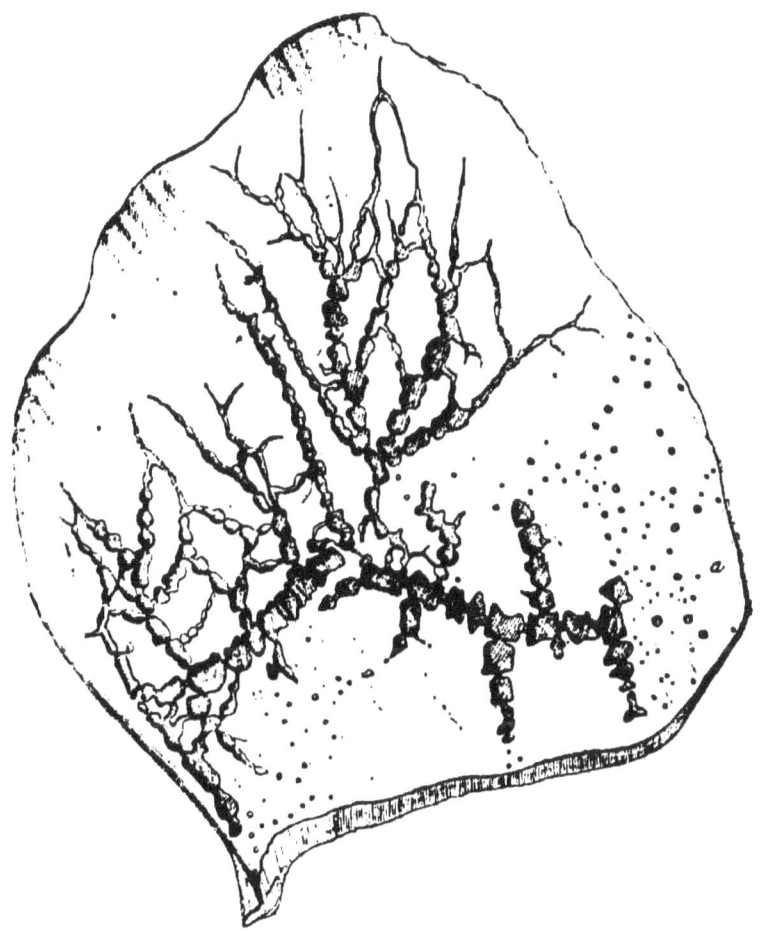

FIG. 70.—BORING SPONGE (*Cliona corallinoides*) IN SHELL.

fied; each species being always found in the same-formed cavities; that is, those with the same kind of spicules, and with papillæ of the same size, number, and arrangement, are always found to branch and to anastomose in a similar manner, and to have the terminal twigs of the same character. This surely could not happen, did *Cliona* take up its abode in cavities caused by decay, or in excavations formed by worms, and were its shape dependent upon such accidental circumstances." The prevailing belief, at that time was, " that *Cliona* does not excavate the chambers in which it is found ; but that they are formed by worms, or by decay, or are produced in some other accidental manner, and that the shape of the sponge depends on that of the cavities it may chance to inhabit." In opposition to which, he continues, " It is pretty evident that they must form their own habitation," and this he illustrates by reference to a particular species, in which " the principal stems take a zigzag direction, sending off at the angles lateral branches, which pass on to unite with the neighbouring stems : the terminal twigs are delicate and bifurcate, one of the divisions going immediately to form its junction with the adjoining stem. This mode of growth goes on, until the entire substance of the shell, in which the *Cliona* is lodged, is completely filled with a network of branches ; the anastomosing increasing all the while, by the addition of twigs from the main stems, until very little of the shell is left to separate the various parts of the sponge. Now, in all this there is nothing having the appearance of accident. Where the *Cliona* is not, the shell is perfectly sound and untouched; the terminal twigs are all alike delicate, and of similar character, penetrating the hard, perfect substance, the main stems become gradually and proportionately thick, and the anastomosis, though somewhat irregular, is identical throughout.

If the sponges are incapable of excavating the chambers in which they conceal themselves, it is inquired how we shall account for the dendritic cavities. "They are evidently not the result of decay, neither are they the burrows of worms; which, when in shell, or other hard calcareous substance, are always linear, sometimes cylindrical, often depressed, never lobed, and frequently double, that is, with two channels, divided by an elevated ridge. And so different are they in their general appearance, that it is very easy to point out which is the excavation of the worm, and which that of the *Cliona*, when the burrows of the two interfere with each other, which not unfrequently occurs."

This writer then proceeds to point out that the walls of the cavities, inhabited by the sponge, are distinctly punctured in a peculiar manner, which must be caused by the character of the surface of the inhabitant. So' certain a test is this stated to be, that by it alone the nature of the excavations in fossil shells may be determined with the greatest confidence.[1]

Much more recently, H. J. Carter says of a species of *Cliona*, that not only does it excavate shells, but the sandstone rock, too, of the same locality where it shelters itself, under the florid expansions of the Nullipores. And in 1881 Mr. B. W. Priest, who had previously held an opposite view, frankly declared that "he had come to the conclusion that no known worm could bore in the manner indicated, where the ramifications are very numerous, and in some cases very fine, and most of them more or less filled up with the sponge or its remains." And further, " I still hold to the same opinion, that a certain class of sponges are capable of burrowing or boring cavities in hard and soft substances."[2]

Having given a summary of opinion on the boring capacity of these sponges, we must candidly admit that they have not remained unchallenged, and that there have been, and still are, excellent observers who do not consider that the burrowing can be accomplished by the sponge. Foremost amongst these was the late Dr. Bowerbank, who says, " Some naturalists have promulgated the idea that this sponge has the

[1] "On the Excavating Powers of Sponges," by A. Hancock, in "Annals of Natural History," vol. iii. (1849), p. 321.
[2] "Journal of Quekett Microscopical Club," vol. vi. p. 269.

power of excavating the canals, and other spaces, which it usually occupies. My own intimate knowledge of the species has led me to a contrary conclusion. When located in oyster, or other, shells it usually fills entirely the cavities between the two surfaces, but when the canals excavated in the limestones extend to the depth of two or more inches, it frequently occurs that the sponge terminates at the depth of less than an inch, and the remaining part of the canal is quite empty and clean, without the slightest indication of having been ever occupied by sponge; and in one of these perforated stones from Tenby, which I broke through the centre, although it abounded with the sinuous canals, none of them presented the slightest traces of having ever contained sponge; and occasionally, oyster shells, full of perforations, may be found in the same condition. These facts militate strongly against the idea that the excavations are produced by the sponge; and, in addition to them, we must bear in mind that the dermal membrane is quite smooth, and that there are no mechanical appliances, or organs visible, by which such a power of attrition could be exerted."[1] This author contends that the perforations must have been made by worms (annelids), and the cavities subsequently occupied by the sponge.

[1] "A Monograph of the British Spongiadæ," by J. S. Bowerbank (Ray Society, 1866), vol. ii. p. 219.

Another vigorous opponent of the burrowing capacity of sponges is Mr. J. G. Waller, who has criticised the arguments adduced in its favour, and endeavoured to establish not only the improbability, but the impossibility of the operation. He says:—" I have endeavoured to show that the solution of the question is, and must be, in the mode of working the burrows. The markings I have attempted to describe, can be demonstrated to be made by a hard tool, working in the segment of a circle, to which I have drawn attention, as shown by the *Scolytus*. And that such should be mimicked by a sponge, a creature so far down in the scale, would be, if proved, one of the most extraordinary marvels in natural history. It would be altogether without parallel, and it therefore requires the most absolute proofs before it should be accepted. No imaginative dream, no assumption, no jumping to conclusions, because minor points are not understood, can support such a theory, in the face of hard and tangible facts, in full agreement with well-known precedents." And thus he finally sums up by stating his propositions:—" If it can be shown that the sponge is not always in the burrows, even as a whole or a part, there is an end of the theory of a 'boring' sponge. If the limestone burrows at Babbicombe never exhibit traces of the sponge, the theory is also at an end, for no one can doubt but that a similar creature made these. If it can be demonstrated that the burrows are made by a hard, and not a soft, instrument, nor by a solvent, the theory is also at an end, and its further prosecution useless. These propositions I have endeavoured to prove."[1]

Without venturing to determine how the operation is performed, we must confess our inability to recognise its impossibility, and, we may go still further, and intimate our conviction that those who advocate the boring theory have the better of the argument, and that there are stronger presumptions in favour of the boring being accomplished by the sponge, whether the process is evident or not, than in the declaration that it is impossible to take place. It may be a case in which a person is quite justified in suspension of judgment, but the evidence of the opposition is insufficient to acquit the sponge of all complicity in the transaction. Whether the boring is accomplished by annelids or sponges, it is clear that an elaborate system of excavation is performed by some marine organism, and that this animal is deserving of recognition in this place as an excavator.

Mr. Waller, and those who think with him, may say that "boring sponges" is only an hypothesis, but there are some apposite remarks, by one who well

---

[1] "On the so-called Boring or Burrowing Sponge," by J. G. Waller, in "Journal of Quekett Microscopical Club," vol. ii. (1871), p. 269, and vol. vi. (1881), p. 251.

deserves to be heard, on the value of hypotheses in scientific inquiry. Professor Huxley says:—" Do not allow yourselves to be misled by the common notion that an hypothesis is untrustworthy simply because it is an hypothesis. It is often urged, in respect to some scientific conclusion, that, after all, it is only an hypothesis. But what more have we to guide us in nine-tenths of the most important affairs of daily life than hypotheses, and often very ill-based ones? So that in science, where the evidence of an hypothesis is subjected to the most rigid examination, we may rightly pursue the same course. A man may say, if he likes, that the moon is made of green cheese: that is an hypothesis. But another man, who has devoted a great deal of time and attention to the subject, and availed himself of the most powerful telescopes, and the results of the observations of others, declares that in his opinion it is probably composed of materials very similar to those of which our own earth is made up: and that is also only an hypothesis. But, I need not tell you, that there is an enormous difference in the value of the two hypotheses. That one which is based on sound scientific knowledge is sure to have a corresponding value; and that which is a mere hasty random guess is likely to have but little value. Every great step in our progress in discovering causes has been made in exactly the same way. A person observing the occurrence of certain facts, and phenomena, asks naturally enough, what process, what kind of operation known to occur in nature, applied to the particular case, will unravel and explain the mystery? Hence you have the scientific hypothesis; and its value will be proportionate to the care and completeness with which its basis had been tested and verified. It is in these matters, as in the commonest affairs of practical life, the guess of the fool will be folly, while the guess of the wise man will contain wisdom. In all cases, you see that the value of the result depends on the patience and faithfulness with which the investigator applies to his hypothesis every possible kind of verification." [1]

---

[1] "On our Knowledge of the Causes of the Phenomena of Organic Nature," by Professor Huxley (1863), p. 66.

PLATE 1.

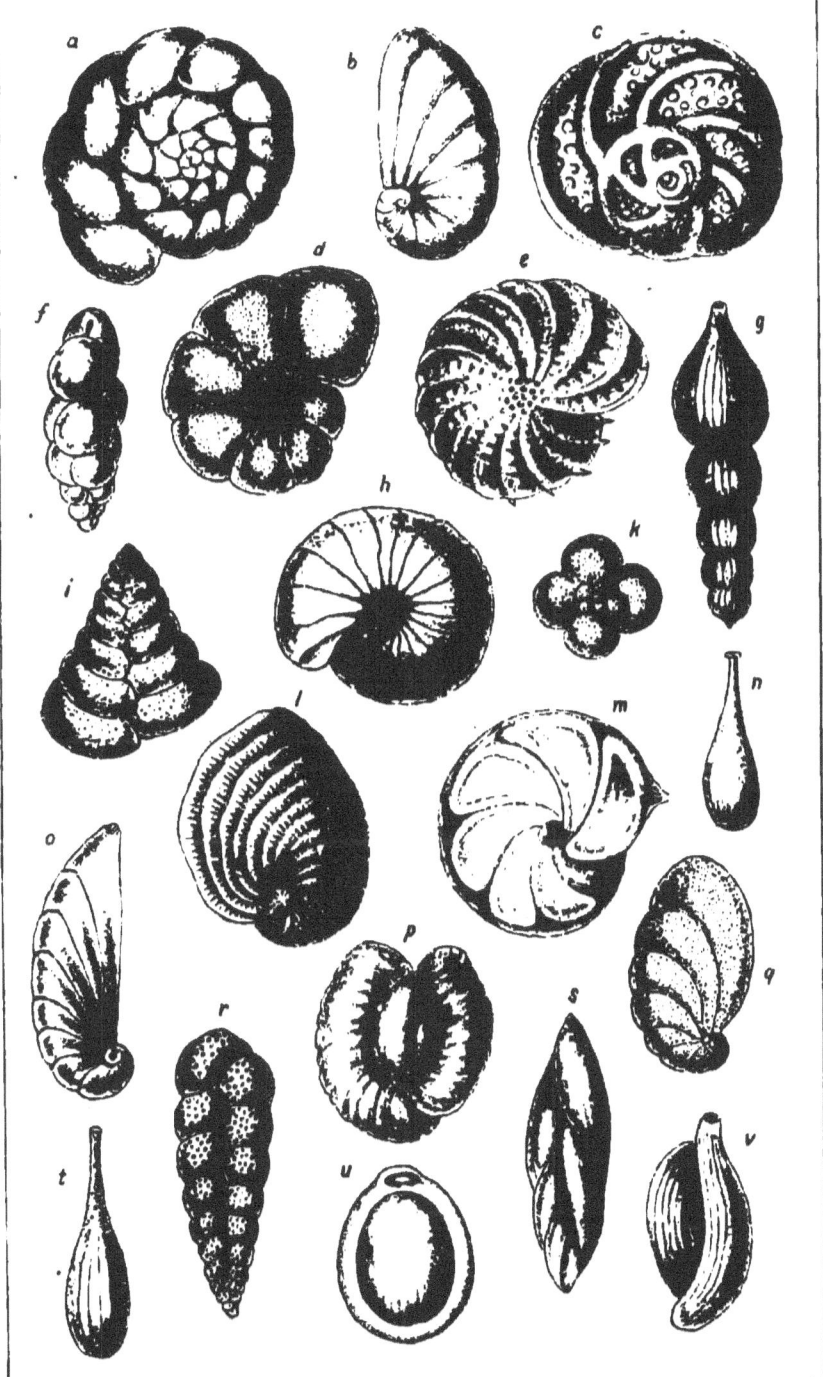

FORAMINIFERA.

PLATE 2.

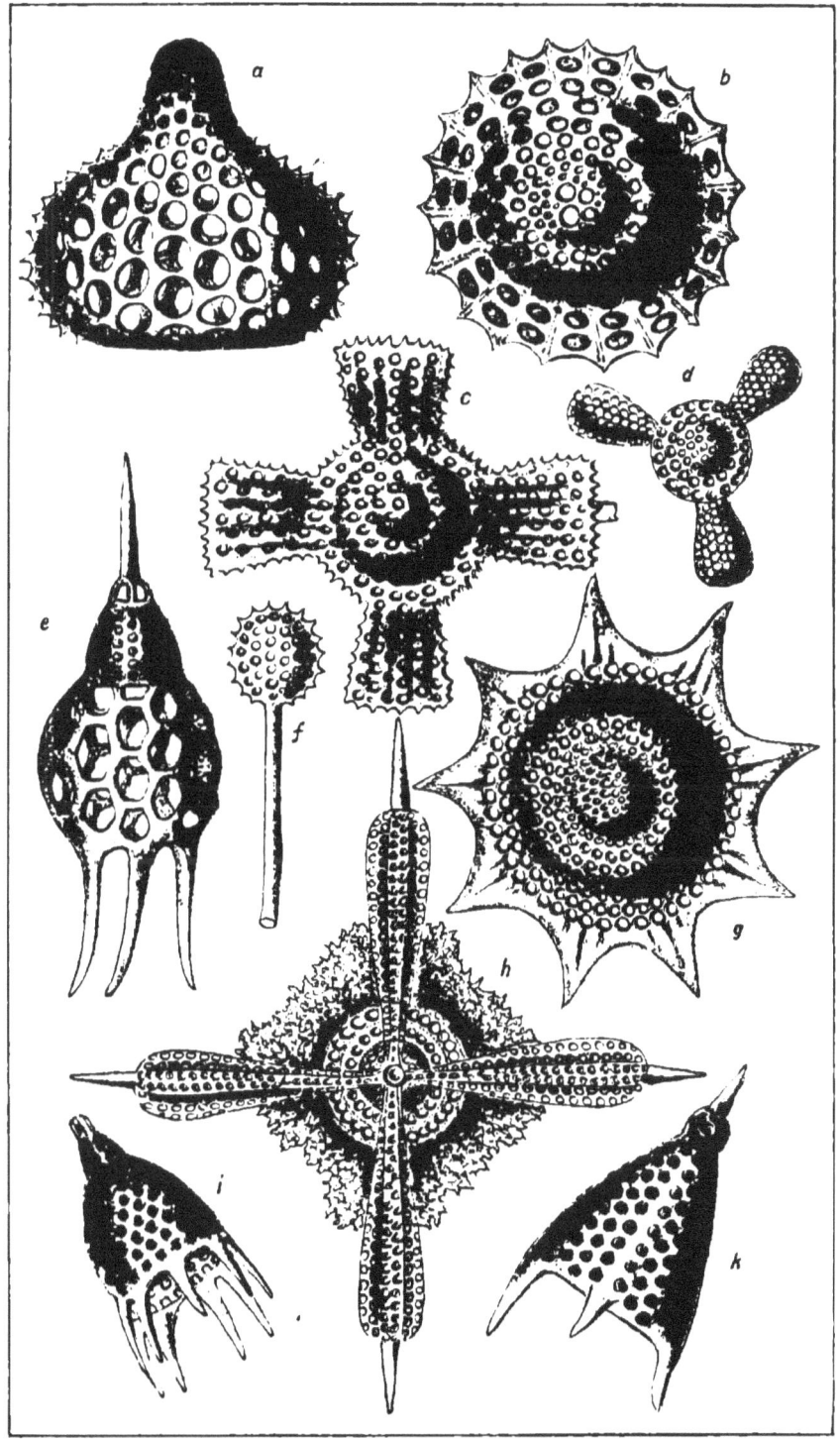

POLYCYSTINS.

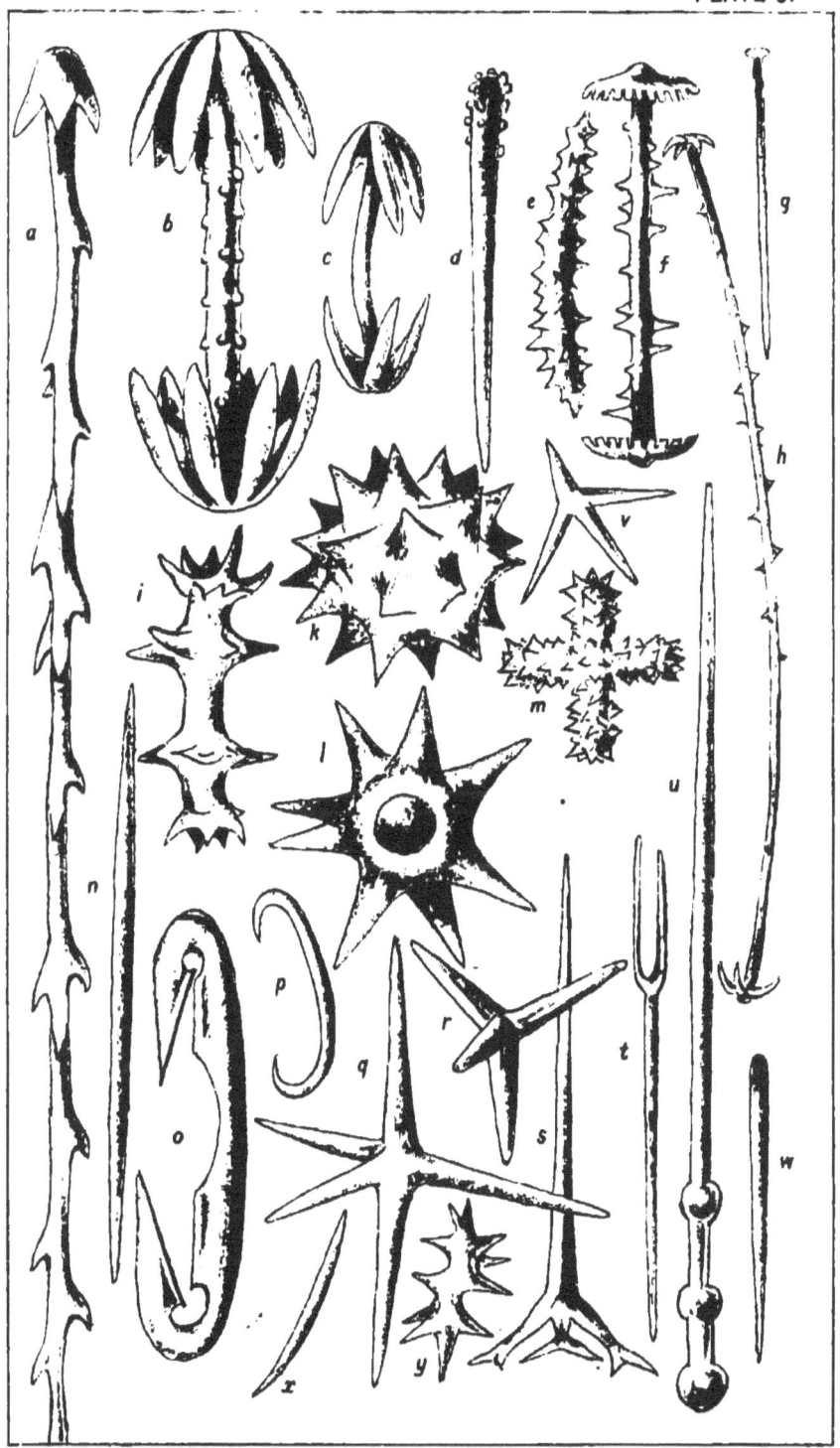

PLATE 3.

SPONGE SPICULES.

PLATE 4.

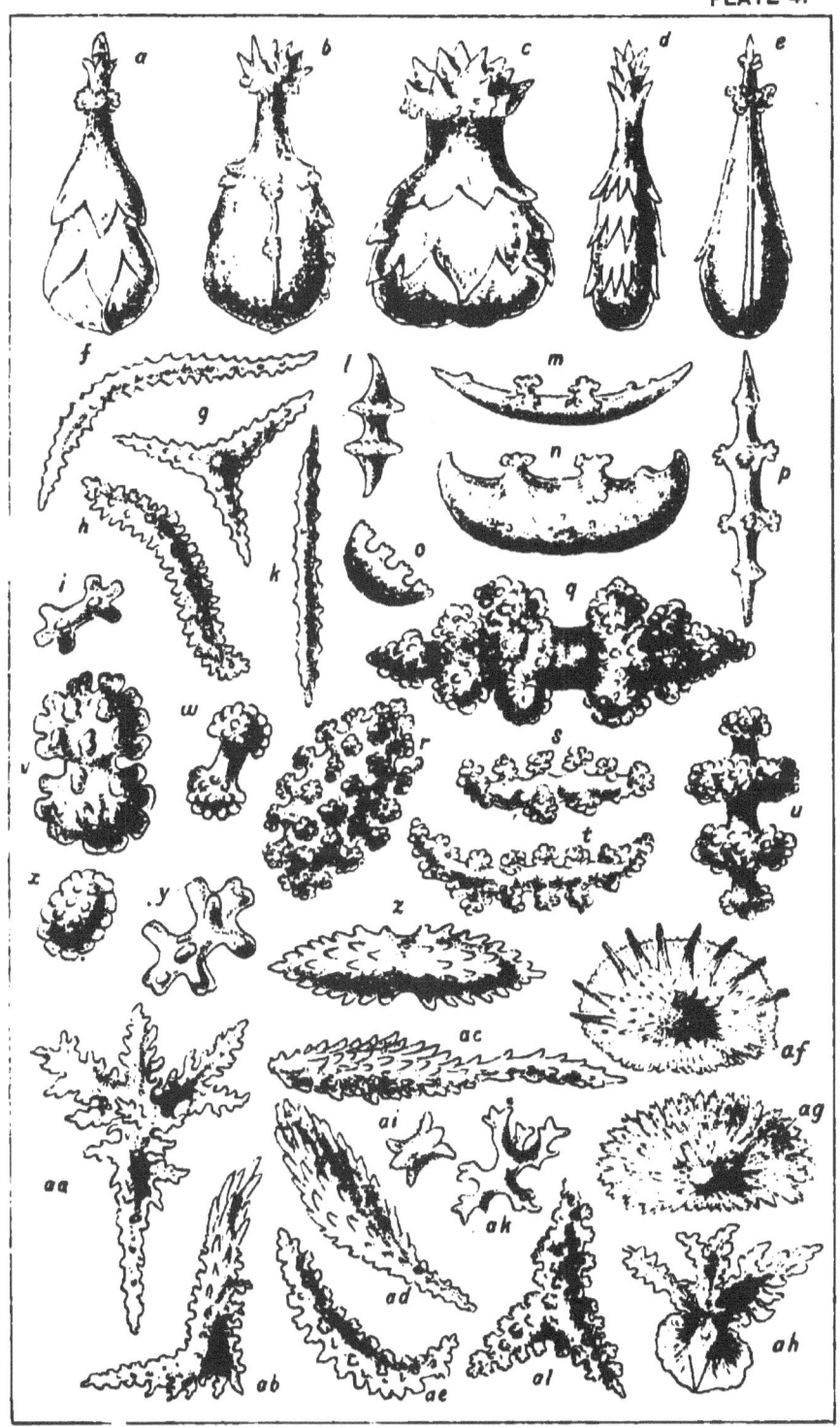

GORGONIA SPICULES.

INDEX.

Absence of light in deep sea, 16
Acontia of sea anemone, 228
Actinophrys, or sun animalcule, 76
Alternation of generations, 97
Amœba, or proteus, 29, 75
Annelids, or tube-making worms, 315
Argyll *versus* Darwin, 277
Astræa corals, 234
Atlantic ooze and chalk, 51
Atolls or Lagoon Islands, 248, 254
Avicularium and vibraculum, 305

Barrier reefs, 248
Basis of life, 25
Bathybius, 21
Bird's head processes, 299
Blood of annelids, 320
Boring of teredo, 343
Boring sponges, 352
Brain corals, 235
"Brown bodies" of polyzoa, 297
Budding of annelids, 318

Caryophyllia Smithii, 231
Chalk makers, or foraminifera, 28
Chylaqueous fluid, 321
Ciliated sponge gemmules, 137
Circulation in radiolaria, 100
Circulatory system of sponge, 128
Cliona, or boring sponge, 352
Cock's-comb or sea-pen, 208
Collared monads, 124
Continuity of the chalk, 53
Coral borers, 242

Coral builders, 215
Coral fishery, 204
Coral, growth of, 266
Coral islands, Darwin's theory, 272
Coral messmates, 316
Coral, organ-pipe, 190
Coral, red, 196
Coral reefs and islands, 247
Coral reefs, distribution of, 260
Coral reefs, thickness of, 251
Corals and limestone, 240
Corals, gemmation of, 232
Cow's paps, or dead man's toes, 180
Cyclosis, or streaming of granules, 83

Darwin's theory of coral islands, 272
Deep sea boring annelids, 347
Deep sea bottom, 17
Deposit of shell structure, 90
Depths of living coral, 263
Depths of the sea, 7
Difflugia, 34
Diffusion of foraminifera, 62
Discovery of polycystins, 69
Distribution of coral reefs, 260
Distribution of radiolaria, 103

Excavating crustaceans, 349
Excavating sponges, 360
Excavators, 340

Fishery of red coral, 203
Fixed polypes, 156
Flagellate monads of sponge, 121

2 B 2

Flustra, or sea-mat, 310
Food of sponges, 130
Foraminifera, 28
Foraminifera and chalk, 55
Foraminifera, diffusion of, 62
Foraminifera, movements of, 38
Foraminifera of the deep sea, 51
Foraminifera of the past, 63
Foraminifera, reproduction of, 40
Foraminifera, shell structure, 43
Foraminifera, variety of form, 46
Forms of foraminifera, 46
Fossil radiolaria, 105
Fringing reefs, 248

Gemmation in radiolaria, 97
Gemmation in sponges, 134
Gemmation of corals, 232
Gorgonidæ, or sea-fans, 180
Growth of coral, 266
Growth of the polypary, 166

Homes or skeletons, 1
Hydra, fresh water, 151
Hydroid corals, 236
Hydroid zoophytes, 158
Hypotheses, to test, 288

Isospores and anisospores, 98

Lagoon islands or atolls, 248, 254
Lasso cells, or stinging cells, 223
Lattice workers, or polycystina, 66
Life history of sabella, 329
Limestone and corals, 240

Medusa-forms, 160
Millepores, zooids of, 237
Minute shells in chalk, 55
Movements of foraminifera, 38

New colonies of zoophytes, 159

Operculum of serpula, 337
Operculum of zoophytes, 153
Organ-pipe coral, 190

Pholas, or boring mollusc, 345
Phosphorescence in deep sea, 211
Phosphorescence in radiolaria, 102
Phosphorescence of sea-pens, 209
Phosphorescence of zoophytes, 169
Plant animals or zoophytes, 145
Polycystina, or lattice workers, 66
Polycystina, reproduction of, 84
Polycystina, variable forms, 71
Polypary, growth of, 166
Polyzoa, or bryozoa, 291
Polyzoa, reproduction of, 307
Polyzoa, structure of, 294
Polyzoon, typical, 293
Pressure at great depths, 13
Pseudopodia, or false feet, 38, 76

Radiolaria and polycystina, 92
Radiolaria, circulation in, 100
Radiolaria, distribution of, 103
Radiolaria, gemmation in, 97
Radiolaria, genera and species, 91
Radiolaria, nutrition of, 99
Radiolaria, phosphorescence in, 102
Radiolaria, skeleton of, 106
Radiolaria, structure of, 92
Radiolaria, zoospores, 96
Rapid growth of zoophytes, 173
Red coral, 196
Reproduction in actinophrys, 80
Reproduction in foraminifera, 40
Reproduction in polycystina, 84
Reproduction in sea anemones, 229
Reproduction in sea-mats, 307
Reproduction in sponges, 132
Reproduction in zoophytes, 164
Reproduction of red coral, 198
Rhizopods feeding, 81

Sabella and serpula, 327
Sarcode or protoplasm, 37
Sarcode, what is it? 74, 120
"Sarcoblasts" of Wallich, 86
Sea anemone, reproduction of, 229
Sea anemone, stinging cells, 228
Sea anemone structure, 219
Sea-fan spicules, 185

INDEX. 373

Sea-fan makers, 180
Sea-mat makers, 291
Sea-pens, or pennatulæ, 207
Shells of foraminifera, 43
Shell structure in foraminifera, 43
Skeleton deposits, 90
Skeleton of radiolaria, 94, 106
Skeleton of red coral, 202
Skeleton of sponges, 112, 118
Skeletons of zoophytes, 147
Spicules of gorgonia, 185
Spicules of sponge, 115, 142
Spirorbis on sea-weed, 338
Sponge, flagellate monads, 121
Sponge gemmation, 134
Sponge, nutriment of, 130
Sponge reproduction, 132
Sponge skeleton, 112, 118
Sponge spicules, 115, 142
Sponge, structure of, 127
Sponge weavers, 109
Sponge, what is it? 109
Sponges, boring or excavating, 352
Sponges, forms of, 111
Stinging organs, or lasso-cells, 223
Streaming of granules, 82
Structure of polyzoa, 294
Structure of radiolaria, 92
Study of marine life, 4
Subsidence, theory of, 272

Temperature of the deep sea, 14
Teredo, or ship-worm, 341
Theories of coral atolls, 269
Thread capsules, or lasso-cells, 223
Tube masons, 314
Tubicolous annelids, 315
Tubiporine, or organ-pipe coral, 190
Type of coral builders, 219
Type of corals, 231

Vegetation of Lagoon Islands, 257
Venus's flower-basket, 111, 139

Weaver tube-worm, 324
Worms! worms! 339

Xanthellæ or yellow bodies, 94, 99

"Yellow bodies" of rhizopods, 85, 94, 99

Zoæcium, or polyp home, 294
Zooids of millepores, 237
Zoospores of radiolaria, 96
Zoophytes, or plant animals, 145
Zoophytes, medusa-forms, 160
Zoophytes, new colonies, 159
Zoophytes, phosphorescence, 169
Zoophytes, rapid growth, 173
Zoophytes, reproduction, 164
Zoophytes, to kill, 154

THE END.

www.ingramcontent.com/pod-product-compliance
Lightning Source LLC
Chambersburg PA
CBHW032025220426
43664CB00006B/364